SILICON NITRIDE—1

SILICON NITRIDE—1

Edited by

SHIGEYUKI SŌMIYA
Nishi-Tokyo University, Japan

MAMORU MITOMO
National Institute for Research in Inorganic Materials, Ibaraki, Japan

MASAHIRO YOSHIMURA
Tokyo Institute of Technology, Yokohama, Japan

Ceramic Research and Development in Japan—Vol. 1

ELSEVIER APPLIED SCIENCE
LONDON and NEW YORK

ELSEVIER SCIENCE PUBLISHERS LTD
Crown House, Linton Road, Barking, Essex IG11 8JU, England

Sole Distributor in the USA and Canada
ELSEVIER SCIENCE PUBLISHING CO., INC.
655 Avenue of the Americas, New York, NY 10010, USA

WITH 41 TABLES AND 90 ILLUSTRATIONS

ENGLISH LANGUAGE EDITION
© 1990 ELSEVIER SCIENCE PUBLISHERS LTD

This is the English-language version of
Chikka Keiso Senamikkusu
© 1987 Shigeyuki Sōmiya and Ūchida Rokakuho Publishing Co. Ltd.

British Library Cataloguing in Publication Data

Silicon Nitride—V.1.—(Ceramics Research in Japan)
 1. Ceramics. Raw materials: Silicon nitrides
 I. Sōmiya, Shigeyuki II. Series
 III. Chikka keiso senamikkusu
 English 666

 ISBN 1-85166-329-0

Library of Congress Cataloging-in-Publication Data

Chikka keiso senamikkusu. English.
 Silicon nitride/edited by Shigeyuki Sōmiya.
  p. cm.
 Bibliography: p.
 Includes index.
 ISBN 1-85166-329-0
 1. Silicon nitride—Congresses. 2. Ceramics—Congresses.
 I. Title
 TP245.N8C4713 1989
 666—dc 19

No responsibility is assumed by the Publisher for any injury and/or damage to persons or property as a matter of products liability, negligence or otherwise, or from any use or operation of any methods, products, instructions or ideas contained in the material herein.

Special regulations for readers in the USA

This publication has been registered with the Copyright Clearance Center Inc. (CCC), Salem, Massachusetts. Information can be obtained from the CCC about conditions under which photocopies of parts of this publication may be made in the USA. All other copyright questions, including photocopying outside of the USA, should be referred to the publisher.

All rights reserved. No part of this publication may be reproduced, stored in a retrieval system, or transmitted in any form or by any means, electronic, mechanical, photocopying, recording, or otherwise, without the prior written permission of the publisher.

Photoset and printed in Northern Ireland at The Universities Press (Belfast) Ltd.

*To
Professor Emeritus Shinroku Saito,
Tokyo Institute of Technology
and
Professor Dr Günter Petzow,
Max Planck Institute für Metallforschung,
Pulvermetallgisches Laboratorium.
They encouraged the study of Si_3N_4 as early as 1960.*

Preface

Silicon nitride ceramics are receiving considerable research attention because they have a useful range of characteristics: strong for thermal shock; high strength at high temperatures; corrosion resistance against fused metals, especially fused non-ferrous metals; chemical inertness; oxidation resistance at high temperatures in air; and are therefore finding increasing application as engineering components, e.g. turbocharger rotors, rocker arm chips, bearings, etc. This book provides a view of the art, science and technology of silicon nitride ceramics and is the translation from the Japanese of the proceedings of a meeting held in December 1986. One paper from that meeting, by S. Sōmiya, M. Yoshimura and J. Kase, has not been included because it had been given in English at the Lubeck meeting on Ceramic Materials and Components for Engines.

We hope that the publication of this English language edition will go far in promoting the understanding and development of silicon nitride ceramics.

SHIGEYUKI SŌMIYA
MAMORU MITOMO
MASAHIRO YOSHIMURA

Acknowledgements

We would like to express our appreciation for organizing the meeting to the following:

Faculty member of Research Laboratory of Engineering Materials, Tokyo Institute of Technology, Administrative office of the Institute, invited speakers, chair professor and participants of the meeting.

The following companies made financial contributions:
    Device Center, Hitachi Ltd.
    Nippon Soda Co., Ltd.
    Kureha Chemical Industries Co., Ltd.
    Fundamental Laboratories, NEC Co., Ltd.
    Kawasaki Rozai Co., Ltd.
    Chichibu Cement Co., Ltd.
    Morimura Brothers Co., Ltd.
    Nanko Kemmazai Co., Ltd.
    Nippon Kenkyu Kaihatsu Co., Ltd.
    Onoda Cement Co., Ltd.
    KDK Co., Ltd.
    Matsushita Electric Industries Co., Ltd.

Without their support, we would not have been able to organize the meeting.

We would like to show appreciation to Mr Satoru Uchida for his help with the publication of the Japanese edition, and the staff of Elsevier Science Publishers in London and Tokyo for their help with the publication of the English edition.

Contents

Preface   vii

Acknowledgements   ix

List of Contributors   xv

1 Thermodynamics, phase relations, and sintering aids of silicon nitride
M. Mitomo

1.1	Introduction	1
1.2	Thermodynamics of silicon nitride	2
1.3	Sintering aids of silicon nitride	7
References		11

2 Onoda silicon nitride powders
T. Tsutsumi

2.1	Introduction	13
2.2	Fabrication method	14
2.3	Silicon nitride raw material powder properties	17
2.4	Onoda silicon nitride raw material powders	20
2.5	Summary	22
References		23

3 NKK's silicon nitride powders: SIN series
K. Nakagawa & M. Kato

3.1	Introduction	25
3.2	Production process	27
3.3	Powder characteristics	28
3.4	Slurry, molding, and sintering characteristics	30
3.5	Summary	37

4 Characterization and synthetic process of Si_3N_4 material powders
M. Nakamura, Y. Kuranari & Y. Imamura

4.1	Introduction	40
4.2	Silicon nitride raw material powders	40
4.3	Synthesis of silicon nitride raw material powders using the metallic silicon method	44
4.4	Synthesis of silicon nitride raw material powders using the silicon chloride and ammonia method	48
4.5	Characteristic comparison of powders obtained from metallic silicon and silicon chloride ammonium methods	51
4.6	Summary	57
	Acknowledgments	57
	References	57

5 α-Si_3N_4 powder produced by nitriding silica using carbothermal reduction
T. Ishii, A. Sano & I. Imai

5.1	Introduction	59
5.2	An outline of silica reduction nitriding synthesis	60
5.3	Characteristics of powders synthesized using carbothermal reduction	66
5.4	Summary	68
	References	69

6 Developments in Si_3N_4 powder prepared by the imide decomposition method
Y. Kohtoku

6.1	Introduction	71
6.2	Experiments	74
6.3	Results and discussion	75
6.4	Summary	80
	References	80

7 State of the art of silicon nitride powders obtained by thermal decomposition of $Si(NH)_2$ and the injection molding thereof
T. Arakawa

7.1	Introduction	81
7.2	The imide thermal decomposition method and characteristics of powders produced thereby	82
7.3	Injection molding properties of imide thermal decomposition powders	86
7.4	Summary	90
	References	91

8 Synthesis of ultrafine Si$_3$N$_4$ powder using the plasma process and powder characterization
N. Kubo, S. Futaki & K. Shiraishi

8.1	Introduction	93
8.2	Experiments	94
8.3	Results and discussion	96
8.4	Summary	104
	References	105

9 The influence of Si$_3$N$_4$ powder characteristics on sintering behavior
K. Ichikawa

9.1	Introduction	107
9.2	Experimental method	108
9.3	Results and discussion	110
9.4	Summary	115
	References	116

10 Properties and applications of Si$_3$N$_4$ whiskers
K. Niwano

10.1	Introduction	117
10.2	Fabrication and characteristics of silicon nitride whiskers	118
10.3	Applications of silicon nitride whiskers	123
10.4	Conclusions	129
	References	130

11 Joining of Si$_3$N$_4$
N. Iwamoto

11.1	Introduction	131
11.2	Trends in joining research	132
11.3	Reaction of Si$_3$N$_4$ and various metals	135
11.4	Compatibility of Si$_3$N$_4$ and alloys	141
11.5	The foundation for silicide formation	141
11.6	Problems on the metal side	146
11.7	Si$_3$N$_4$–Si$_3$N$_4$ joining using oxide solder	147
11.8	Summary	150
	References	151

Index 157

List of Contributors

T. ARAKAWA
Advanced Materials Research Laboratories, TOSOH Co., 2743-1 Hayakawa, Ayase, Kanagawa 252, Japan

S. FUTAKI
Central Research Laboratory, Sumitomo Metal Mining Co. Ltd, 3-18-5 Nakakokubun, Ichikawa, Chiba 272, Japan

K. ICHIKAWA
Shiojiri Research Laboratory, Showa Denko K.K. 1, Soga, Shiojiri, Nagano 399-64, Japan

I. IMAI
Toshiba Ceramics Co. Ltd, Research & Development Center, Ogakie, Kariya, Aichi 448, Japan

Y. IMAMURA
Omuta Plant Research and Development Division, Denki Kagaku Kogyo Co. Ltd, Omuta, Fukuoka 836, Japan

T. ISHII
Toshiba Ceramics Co. Ltd, Research & Development Center, 30, Soya, Hadamo-City 257, Japan

N. IWAMOTO
Welding Research Institute, Osaka University, 11-1 Mihagaoka, Ibaraki, Osaka 567, Japan

List of Contributors

M. KATO
Technology Section, Toyama Works, NKK Corporation (Nippon Kokan), Shinminato, Toyama 934, Japan

Y. KOHTOKU
New Ceramic Materials Research Group, Ube Laboratory, Corporate Research & Development, Ube Industries Ltd, 1978-5 Kogushi, Ube, Yamaguchi 755, Japan

N. KUBO
Central Research Laboratory, Sumitomo Metal Mining Co. Ltd, 3-18-5 Nakakokubun, Ichikawa, Chiba 272, Japan

Y. KURANARI
Performance Materials Department, Denki Kagaku Kogyo Co. Ltd, Chiyoda, Tokyo 100, Japan

M. MITOMO
National Institute for Research in Inorganic Materials, Science and Technology Agency, 1-1 Namiki, Tsukuba-shi, Ibaraki 305, Japan

K. NAKAGAWA
Technology Section, Toyama Works, NKK Corporation (Nippon Kokan), Shinminato, Toyama 934, Japan

M. NAKAMURA
Omuta Plant Research and Development Division, Denki Kagaku Kogyo Co. Ltd, Omuta, Fukuoka 836, Japan

K. NIWANO
Tateho Chemical Industries Co. Ltd, 974 Kariya, Ako, Hyogo 678-02, Japan

A. SANO
Toshiba Ceramics Co. Ltd, Research & Development Center, Ogakie, Kariya, Aichi 448, Japan

K. SHIRAISHI
Central Research Laboratory, Sumitomo Metal Mining Co. Ltd, 3-18-5 Nakakokubun, Ichikawa, Chiba 272, Japan

T. TSUTSUMI
Ceramics R & D Laboratory, Onoda Cement Co. Ltd, 1-1-7 Toyosu, Koto-ku, Tokyo 135, Japan

1 | Thermodynamics, Phase Relations, and Sintering Aids of Silicon Nitride

M. MITOMO

ABSTRACT

Stability relations of α- and β-Si_3N_4, Si_2N_2O, SiO_2, and Si in the system Si–O–N are discussed in relation to thermodynamic functions and formation conditions. It is shown that Si_3N_4 is stable under very low oxygen partial pressures.

The α to β transformation in Si_3N_4 is based on a reconstructive one, i.e., by diffusion through a vapor or liquid phase. The effect of additive on the microstructure of sintered materials is also shown by phase relations containing Si_3N_4 and an additive.

1.1 INTRODUCTION

The two components of silicon nitride, silicon and nitrogen, are abundant on the Earth's surface and in the atmosphere. Silicon nitride itself, however, does not occur naturally. Consequently, a non-oxidizing atmosphere is required for the synthesis of silicon nitride powder and the preparation of sintered bodies. The relation between the characteristics, sinterability of the starting powders and the properties of the sintered body are not sufficiently understood. This chapter presents a summary, which is generally applicable to any type of powder, of the role of sintering aids and their effect on the microstructures.

1.2 THERMODYNAMICS OF SILICON NITRIDE

1.2.1 Phase Diagram for the Si–N–O System

This system consists of Si, SiO_2, Si_2N_2O, and Si_3N_4. It is possible to derive the boundaries of each phase from the thermodynamic function of each material. Thermodynamic functions of Si_2N_2O and Si_3N_4 have been determined by measuring the pressure of the vapor phase at equilibrium with the solid, or measuring the thermal decomposition pressure of the solid. Blegan[1] determined the free energy for the formation of β-Si_3N_4 by measuring the relationship between temperature and the equilibrium nitrogen pressure in the reaction between Si (in an Fe–Si melt) and nitrogen:

$$3Si(Fe-Si) + 2N_2(g) \rightleftarrows Si_3N_4(s) \tag{1}$$

$$\Delta G^0_{(Si_3N_4)} = RT \ln(P^2_{N_2} \cdot a^3_{Si}) \tag{2}$$

In the same way:

$$3Si(Fe-Si) + SiO_2(s) + 2N_2(g) = 2Si_2N_2O(s) \tag{3}$$

$$\Delta G^0_{(Si_2N_2O)} = RT \ln(P_{N_2} \cdot a^{3/2}_{Si}) + \tfrac{1}{2}\Delta G^0_{(SiO_2)} \tag{4}$$

The pressure of the vapor phase at equilibrium with the solid was also measured, using a Knudsen cell. From these data it is possible to calculate the stable region of P_{N_2} and P_{O_2} for each phase at 1600–1700 K. The results are shown in Fig. 1.1. The α-Si_3N_4 in Fig. 1.1 is not a true nitride, but rather the oxynitride represented by $Si_{23}N_{30}O$, with a range located between β-Si_3N_4 and Si_2N_2O.

As is clear from the figure, silicon nitride is stable at nitrogen pressures of more than 10^{-2} atm and oxygen pressures of less than 10^{-20} atm. It can thus be seen that silicon nitride has thermodynamic stability only at extremely low oxygen pressures. When the actual conditions under which silicon nitride is synthesized are examined, however, it is interesting to note that in most cases these stable regions are not utilized. For example, when powder is synthesized using the nitridation reaction of Si,

$$3Si + 2N_2 \rightarrow Si_3N_4 \tag{5}$$

the concentration of oxygen in the nitrogen gas is at best 1–10 ppm. Consequently, the nitriding of Si is performed under

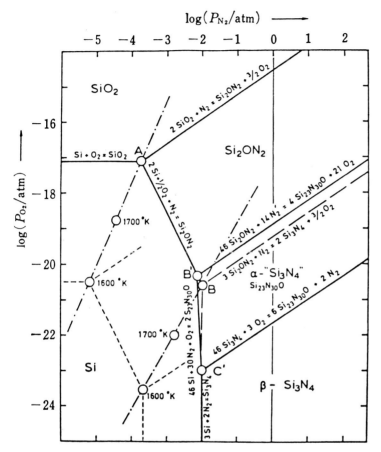

Fig. 1.1. Calculated phase diagram for the Si–O–N system.[1]

thermodynamically stable conditions for SiO_2. Since the actual amount of oxygen in the atmosphere is very small, it is either trapped as SiO_2 or released from the system as SiO. As a result, a significant portion of the material formed will be Si_3N_4. Since the actual reaction may well occur under conditions that are fairly out of equilibrium, a review of the stability relations from experimental results will require considerable caution.

Jack[2] has shown that the stable phases and regions at 1600 K differ from those at 1800 K (see Fig. 1.2). According to these results, β is the only stable form of pure Si_3N_4. Type α, which is stable at low temperatures, is actually an oxynitride. In any case,

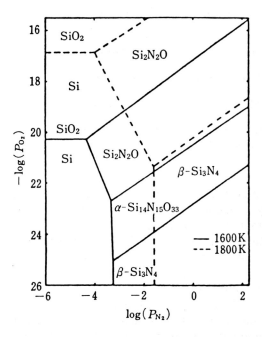

Fig. 1.2. Phase diagram for the Si–O–N system.[2]

silicon nitride is stable only at extremely low oxygen pressures, and it can be safely assumed that it does not occur naturally.

1.2.2 The Relationship Between α and β

A starting powder with a high α content is required for use in sintering, and this in turn makes it necessary to understand the relationship between α and β. No definite conclusions have been reached as yet, but the currently prevailing theories are offered in Table 1.1.

The first explanation proposes that α is a low-temperature form and β a high-temperature form. This is based on the following observations: first, the lower the temperature under which synthesis is performed, the higher the α content of the resulting powder; second, the transformation from α to β occurs at high temperatures. Although this is the most widely accepted explanation, the transformation from β to α has not been observed, and it seems to be impossible to declare with certainty that α is a low-temperature type.

Theory	α	β	
			Table 1.1 Theories Explaining the α–β Relationship
I	Low-temperature	High-temperature	
II	Unstable at any temperature	Stable at any temperature	
III	High-P_{O_2}	Low-P_{O_2}	
IV	Stable in low-temperature, high-P_{O_2} regions	Stable at any temperature	

The second explanation holds that only β has a stable form, α being unstable at any temperature. This is based upon the fact that a β-to-α transformation has never been observed.

According to the third explanation (see Fig. 1.1), the stability of α and β are affected more by oxygen pressure than by temperature, and α is treated as a silicon oxynitride. Doubt has been cast on this explanation, however, by the fact that it has proven possible to obtain pure, low oxygen content α-Si_3N_4 single crystals using chemical vapor deposition.[3,4] The fourth explanation, based largely on recent findings in thermodynamic experiments, falls somewhere between the second and third.[2]

There is as yet no consensus on the relationship between α and β, primarily because of the extremely wide range of phenomena based on a variety of conditions for synthesis and for pseudo-equilibrium states. It is difficult for a single theory to offer a comprehensive explanation.

Figure 1.3[2] is helpful in understanding the relationship between α and β structures. The β structure consists of stacked layers AB, shown to the upper left of the figure, while in the α

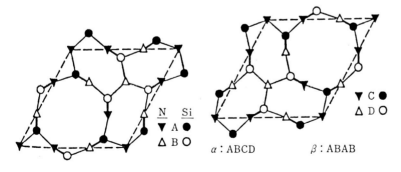

Fig. 1.3. Unit layers AB (left) and CD (right) for silicon nitride.

structure the layer CD, shown to the right, is added to form stacked layers ABCD. At a glance, the AB and CD layers appear to have twofold rotational axes. Since it seems possible for the α structure to be transformed into β through rotation, it was mistakenly believed that a phase transformation was occurring within the solid. It is possible, for example, for 2H SiC (wurtzite) to transform to 3C (zinc-blende structure) when a c-axis-related rotation is provided in every three layers. In fact, when 2H whiskers are heated, a structure similar to 3C will appear in the form of stacking faults, and the whiskers will ultimately develop into 3C whiskers. This type of solid-state transformation is a unique phenomenon observed only in whiskers. The SiC polytypic phase changes which occur in powder develop due to diffusion through the vapor or liquid phase and do not show any solid-state transformation. This is believed to be because the activated energy is too large in such phase transformations.

Layers AB and CD in Fig. 1.3 are actually mirror images of one another, and they do not coincide by rotation. As a result, a phase change from α to β or from β to α can be achieved only by breaking the Si–N bond. The phase change from α to β that has been observed occurs only as the result of diffusion through the vapor or liquid phase. It is possible to conclude that the absence of a documented β-to-α transformation has been because temperatures have been too low and diffusion through the vapor or liquid phase has been too slow.

1.2.3 Preferred Raw Powders

Numerous characteristics are required for powders to be used in sintering. They include: (1) high purity; (2) fine grain (0·1–0·5 μm); (3) free of agglomerates; (4) narrow particle size distribution; (5) spherical shape; and (6) low cost. Numerous methods capable of synthesizing high-purity raw materials at low temperatures have been developed. These methods utilize conditions that facilitate the formation of α both thermodynamically and kinetically. In those powders containing both α and β, α forms on the surfaces of large particles or as small particles. As a result, powders with a high α-phase are characterized by a fine grain and a large surface area.

In the past, powders with a high α content were preferred because of ease of sintering, as well as the resulting mechanical characteristics of the sintered body. This is due less to substantial differences in the crystal phase, however, than to differences in powder characteristics, which in turn are based upon differences in synthesis conditions. In the synthesis methods developed thus far, for example, β synthesis requires high temperatures and produces a large particle. Consequently, a high α content would not be required for a raw material powder.

1.3 SINTERING AIDS OF SILICON NITRIDE

1.3.1 Characteristics of Sintering Aids

Most sintering aids are oxides that react with silicon nitride or its oxide layer (silica) to produce a liquid phase at sintering temperatures. The sintering of silicon nitride is performed by the rearrangement and solution–reprecipitation processes, both of which are brought about by the liquid phase. Three important characteristics for a sintering aid are: (1) stability at high temperatures; (2) the existence of a region that produces a liquid phase in the Si_3N_4–SiO_2-oxide system; and (3) solubility of Si_3N_4 in the resulting liquid phase.

Negita[5] classified various oxides according to their effectiveness in sintering, and proposed that both the free energy of formation and the melting point must be above a specified value (see Fig. 1.4). This corresponds with the above condition that the oxide must have high-temperature stability.

The main process in sintering is that of solution–reprecipitation. In this step, the material dissolves in the liquid phase and, after diffusion, precipitates in low-energy positions. At the same time, crystal growth is occurring due to the α-to-β transformation and the solution of small particles, and their precipitation on large particles. Since silicon nitride ceramics are commonly used in structural components, mechanical characteristics such as strength, thermal shock resistance, and fracture toughness are very important. As mechanical characteristics are determined by the microstructure of the ceramic in question, the object of sintering is not simply to achieve a high density but to produce a desired microstructure.

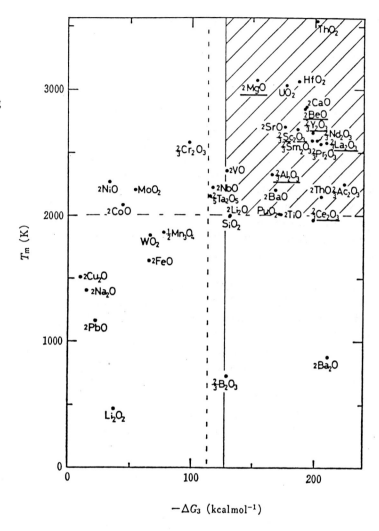

Fig. 1.4. The range of the melting point (T_m) and the free energy of formation (ΔG_3) for effective sintering aid oxides.

When the ceramic is cooled after sintering, the liquid phase will solidify into a glassy phase. In silicon nitride ceramics, therefore, a glass phase remains, at least to some extent, at the grain boundaries. The condition of the majority of the sintering aids has a significant influence on the mechanical and thermal characteristics of the ceramic material.

1.3.2 Classification of Sintering Aids

1.3.2.1 Materials Remaining at the Grain Boundaries After Sintering

(1) Formation of a glassy phase at the grain boundaries: MgO. Use of MgO as a sintering aid produces a relation like the one shown in Fig. 1.5. The sintering body has Si_3N_4 particles as a main component and a liquid phase of Mg–Si–O–N system (shaded region). The composition of the liquid is in equilibrium with the Si_3N_4. The alkali metals and alkali earth metals within the raw material powder are also segregated within this liquid phase. Since the softening point of this type of glass is quite low, it is unlikely that a ceramic with good high-temperature strength could be produced using MgO.

(2) Formation of a crystalline phase: Y_2O_3. When Y_2O_3 is used, $Y_2Si_3O_3N_4$ ($Si_3N_4 \cdot Y_2O_3$) forms at the grain boundaries as a crystalline phase in addition to Si_3N_4 (see Fig. 1.6). A liquid phase with a composition similar to $YSiO_2N$ forms as the third phase. The liquid phase is necessary in order for sintering to proceed, and this is identical to the case in which MgO is used. Since a significant portion of the Y_2O_3 reacts with the Si_3N_4 and crystallizes, the amount of the glassy phase at the grain boundaries is less than in the case of MgO. Furthermore, the glass of the Y–Si–O–N system has a higher softening point than that of

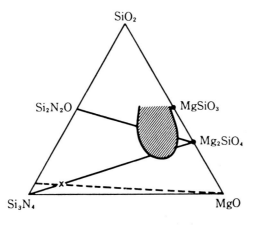

Fig. 1.5. Si_3N_4–SiO_2–MgO phase relations at 1550°C (the shaded portion indicates melting).

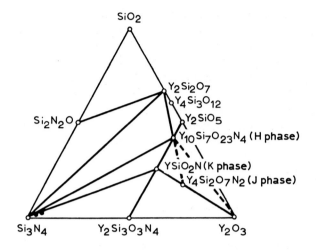

Fig. 1.6. Si_3N_4–SiO_2–Y_2O_3 system phase relations at 1600–1750°C.[6]

the Mg–Si–O–N system. Ceramic materials obtained through the addition of Y_2O_3 can therefore offer better high-temperature strength than when MgO is used.

1.3.2.2 Dissolution Within the Particle After Sintering

When Al_2O_3 is used as a sintering aid, the liquid phase causes densification, with dissolution of the liquid at the same time.

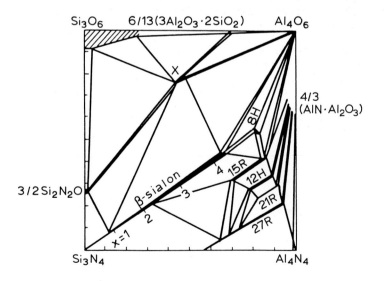

Fig. 1.7. Si_3N_4–SiO_2–Al_2O_3–AlN system phase relations at 1760°C.[7]

Consequently, the ceramic material consists of the Si–Al–O–N system and is referred to as sialon. The relation is as shown in Fig. 1.7.[7] Sialon is a solid solution in which the Si atoms in β-Si_3N_4 are replaced by Al atoms and the N atoms by O atoms. When a Si_3N_4–SiO_2–AlN or Si_3N_4–Al_2O_3–AlN powder with an appropriate mixing ratio is heated, a solid solution develops together with sintering, and the resulting material is a sialon ceramic. Since quite a small amount of the glassy phase remains at the grain boundaries in such materials, they offer excellent resistance to creep and oxidation.[8]

REFERENCES

1. Blegan, K., Equilibrium and kinetics with system Si–N and Si–N–O. In *Special Ceramics*, Vol. 6, ed. P. Popper. *British Ceramic Society*, 1975, pp. 223–38.
2. Jack, K. H., The characterization of α′-sialons and the α–β relationships in sialons and silicon nitride. In *Progress in Nitrogen Ceramics*, ed. F. L. Riley. Martinus Nijhoff, Leyden, 1977, pp. 45–59.
3. Kohatsu, I. & McCauley, J. W., Re-examination of the crystal structure of α-Si_3N_4. *Mat. Res. Bull.*, **9** (1974), 917–20.
4. Kato, K., Inoue, Z., Kijima, K., Kawada, I., Tanaka, H. & Yamane, T., Structural approach to the problem of oxygen content in alpha silicon nitride. *J. Am. Ceram. Soc.*, **58** (1975), 90–91.
5. Negita, K., Effective sintering aids for Si_3N_4 ceramics. *J. Mat. Sci. Lett.*, **4** (1985), 755–8.
6. Lange, F. F., Singhal, S. C. & Kuznicki, R. C., Phase relations and stability studies in the Si_3N_4–SiO_2–Y_2O_3 pseudoternary system. *J. Am. Ceram. Soc.*, **60** (1977), 249–52.
7. Gauckler, L. J., Lukas, H. L. & Petzow, G., Contribution to the phase diagram Si_3N_4–AlN–Al_2O_3–SiO_2. *J. Am. Ceram. Soc.*, **58** (1975), 346–7.
8. Jack, K. H., Sialons and related nitrogen ceramics. *J. Mat. Sci.*, **11** (1976) 1135–58.

2 Onoda Silicon Nitride Powders

T. TSUTSUMI

ABSTRACT

The raw materials for silicon nitride ceramics are produced by several processes. Onoda Cement uses metallic silicon nitriding, which is the most economical method. Control of the reaction temperature during the formation of α-Si_3N_4 with this method is difficult, because the nitriding reaction is an exothermic one, but an original method is used to produce stable α-type silicon nitride. This chapter describes the fabrication method, varieties, and characteristics of Onoda silicon nitride powders.

2.1 INTRODUCTION

Silicon nitride ceramics have recently come to be used in engine parts and other heat-resistant, wear-resistant, or sliding parts, all of which require high performance and high reliability. In order to achieve such characteristics, the silicon nitride raw material powder must be characterized by few irregularities, high purity, fine grain, ease of sintering and good packing properties. Onoda Cement is engaged in the development, manufacture, and marketing of various powders, with the greatest emphasis being placed on the reduction of irregularities through stabilization of the fabrication process.

This chapter describes phenomena confirmed with the company's silicon nitride raw material powders, and introduces fabrication methods, powder varieties, and characteristics.

2.2 FABRICATION METHOD

Onoda Cement produces silicon nitride raw material powders using the direct nitriding of metallic silicon, the most economical method available. The reaction formula is expressed as:

$$3Si + 2N_2 = Si_3N_4$$

The fabrication process employed at Onoda Cement is shown in Fig. 2.1. A detailed description of the process will follow, with reference to the figure.

2.2.1 Adjustment of Metallic Silicon Purity and Particle Size

The particle size of the raw material metallic silicon is brought to a specified value by grinding, and the powder is then mixed for further adjustment of purity. The reason for this will become clear in the description of the baking process, but in essence the reaction rate during synthesis must be stabilized, and this requires the adjustment of particle size and purity (including oxygen). Without close monitoring of the raw material silicon particle size, irregularities also develop in the particle size of the baked body after synthesis, and the process is destabilized resulting in irregularities during the adjustment of product particle size. Adjustment of raw material purity at this stage is critical, since Al in particular forms a solid solution in the silicon nitride, and control of product purity by adjustment in later processes becomes impossible, with the end result being irregularities in product purity.

2.2.2 Baking

In today's market for silicon nitride powders, high-α-content (α-type) powders are favored. The conditions required for synthesis of high-α-content powders using the direct nitriding method have been described in numerous works.[1]

The main problems during actual production, however, are as

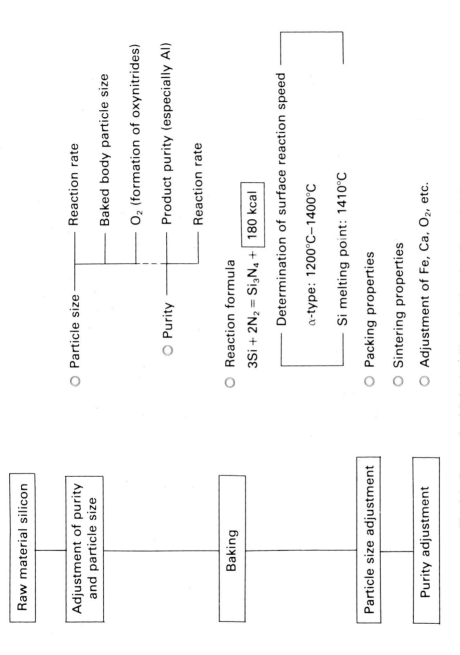

Fig. 2.1. Process for fabricating silicon nitride using direct nitriding.

follows:

    (a) α-type powders are synthesized in the range ~1200–1400°C, while at temperatures above 1400°C β-type tends to be formed.

    (b) This reaction is an exothermic reaction. As a result, it is extremely difficult to control the synthesis temperature of the raw material silicon.

    (c) The reaction is also one in which the surface reaction is rate-determining. Consequently, there is a drastic reaction in the initial phase, and the more raw material silicon present the more difficult it is to control the temperature of the raw material silicon. Moreover, since the melting point of silicon is ~1400°C, the drastic increase in temperature causes melting of the raw material silicon, significantly reducing surface area and making completion of synthesis difficult.

As can be seen from Fig. 2.1, the baking process is the most important in determining the cost of the silicon nitride raw material powder, and differences in baking knowhow make the difference in product cost.[2-4] By focusing on the reaction formula of the direct nitriding method and using an original method of atmospheric control, Onoda Cement has devised and is currently utilizing a method of synthesizing stable $\alpha\text{-Si}_3\text{Na}_4$.

2.2.3 Adjustment of Particle Size

After baking, the particle size of the clinker is adjusted by grinding. This process is critical in determining the material's characteristics as a sintering raw material later on. If foreign substances enter the material during this step or unground grit were to remain, sintered body defects would result directly. The grinding method used also has a major effect on the powder's packing and sintering properties when used as a sintering raw material.

2.2.4 Adjustment of Purity

This process is carried out with the objective of removing impurities originally contained in the material, as well as those introduced during the adjustment of particle size. The purity of the silicon nitride raw material powder is therefore improved.

Using the above processes, Onoda Cement is able to produce and market stable, low-cost silicon nitride raw material powders.

2.3 SILICON NITRIDE RAW MATERIAL POWDER PROPERTIES

Silicon nitride raw material powders are generally sintered using liquid-phase (and occasionally solid-phase) sintering with a sintering aid. As a result, material transfer of silicon nitride during sintering and reactivity with the sintering aid are extremely important. Among the important characteristics used to distinguish silicon nitride sintering raw material powders are sintering and packing properties, and the microstructure and strength of sintered bodies produced using the powder.[5,6]

Materials with a high powder surface activity provide good material transfer and reactivity with the sintering aid, resulting in better sinterability. Good packing properties mean a smaller distance between the silicon nitride molecules and the sintering aid or other silicon nitride molecules, also resulting in better sinterability. Furthermore, since voids are reduced and it is more difficult for pores to remain within the sintered body, the strength and regularity of the sintered body are favorably influenced. The metallic direct nitriding method uses the breakdown technique during grinding, so that the powder surface is activated and sintering properties are improved. In addition, since an optimal particle size distribution can be obtained through adjustment, this method makes possible the fabrication of powders with excellent packing properties. Surface activity in such silicon nitride powders increases with the fineness of the grain. Consequently, oxidation and reduction of the high-temperature strength of the sintered body may occur due to improper pre-sintering handling, while a high agglomeration energy can result in insufficient mixing with the sintering aid and non-uniform sintered body microstructure, or the presence of voids.

At Onoda Cement, a thorough understanding of and control over these phenomena are considered to be important responsibilities in the company's role as a supplier of sintering raw materials. The following is one example of a phenomenon confirmed with Onoda Cement powders.

Wet mixing with sintering aids is often carried out during the fabrication of silicon nitride sintered bodies. Slurry viscosity during mixing is extremely important for the mixing time, the manufacture of granulated powder using spray drying, and cast forming. The following are the results of various measurements conducted on the slurry viscosity of Onoda Cement's SH-5 silicon nitride raw material powder.

2.3.1 Sample Preparation

It is common for silicon nitride raw material powders to be handled in many ways from manufacturing until shipment, and during preservation after shipment to the customer. Harsh treatment of the SH-5 silicon nitride raw material powder was assumed, and the powder was mixed in a closed dry mixer (a rocking mixer developed by Aichi Electric Mfg. Co.) for 30 min, 60 min, and 150 min to prepare samples.

2.3.2 Measurement of Viscosity

Viscosity was measured using a Brookfield-type viscometer after dispersing 150 g of silicon nitride powder in 100 g of water using the mixer. As shown in Fig. 2.2, there was a significant increase

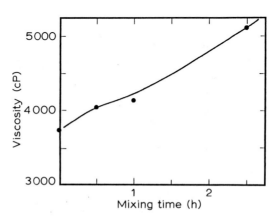

Fig. 2.2. Changes in viscosity.

in viscosity from 3700 cP to 5300 cP after only 150 min of mixing, indicating how easily viscosity can be affected by storage and handling. Since variations in viscosity due to handling are present even within the same production lot, viscosity must be adjusted whenever spray drying or cast molding are performed.

2.3.3 Flock

Observations of slurry showed that variations in viscosity can result in significant gaps in sedimentation speed. After isolating Onoda Cement silicon nitride powders (with various viscosities) with water, the amount of flock larger than 50 μm was measured, and the results are plotted in Fig. 2.3. An extremely high correlation was found between the flock volume and the viscosity: even if the viscosity is adjusted using a high solvent ratio, etc., it is impossible to change the flock volume. Dispersants are normally used in the mixing process, and in some cases the amount of dispersant is used to control viscosity. Since viscosity is changed by leaving the flock alone, it does not become a major problem.

Given this information, water (i.e., adsorption of moisture in the atmosphere) was chosen as a possible reason why the physical characteristics of silicon nitride powders changed with

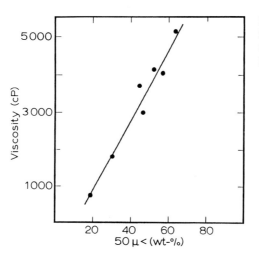

Fig. 2.3. Changes in viscosity of flock larger than 50 μm.

handling. To test this hypothesis, the above-described 5300-cP powder was dried for 10 h at 120°C, and the viscosity and flock volume were measured. A viscosity of 120 cP and a flock volume of less than 10% were shown. When the powder was dry-mixed using the same process, a trend toward increased viscosity and flock was observed. Since the physical characteristics of silicon nitride powders fluctuate with surface conditions, they should be carefully controlled.[7]

2.4 ONODA SILICON NITRIDE RAW MATERIAL POWDERS

This section introduces the silicon nitride raw material powders produced by Onoda Cement. As mentioned earlier, the use of an original baking process allows stable control of the α-content, and most of the powders manufactured and marketed by Onoda Cement have an α-content of 95%.

Some representative Onoda Cement powders are shown in Table 2.1. There are three particle size categories, <1 μm (>95%), <3 μm (>95%), and <5 μm (>95%), and two grades of purity; high purity (in the upper box), and standard purity (in the lower box). Typical chemical compositions for the high-purity SH-3A and the standard-purity HM-5 are shown in Table 2.2. Particle size distributions for four representative Onoda Cement powders are shown in Fig. 2.4. SH-3A has a relatively wide distribution for a high-purity powder and, in fact, was developed for its excellent packing properties.

Table 2.1 Representative Onoda Silicon Nitride Powders (95% α-content)

Purity	Particle size		
	<1 μm (>95%)	<3 μm (>95%)	<5 μm (>95%)
Al < 0·1% Ca < 0·1% Fe < 0·05%	SH-1	SH-3A	SH-5
Al < 0·3% Ca < 0·1% Fe < 0·3%			HM-5

	SH-3A	HM-5
Chemical composition		
Si (%)	58·8	59·0
N (%)	38·9	38·9
Al (%)	0·06	0·25
Ca (%)	0·04	0·03
Fe (%)	0·02	0·21
Mg (%)	0·01	0·02
C (%)	0·1	0·1
O (%)	0·8	1·2
Free Si (%)	>0·1	>0·1
Particle size		
Specific surface area (m^2/g)	11	8
Mean particle size (μm)	0·9	1·4

Table 2.2 Typical Chemical Composition Values for SH-3A and HM-5

Molding properties of the four powders are provided in Table 2.3. The green density values shown are for bodies isostatically pressed at 1 ton/cm^2 after mixing with 6% Y_2O_3 and 4% Al_2O_3, while the oil absorption figures on the right (the amount of oil required to knead 100 g of powder into one lump) are used as a measure of molding properties: the smaller the value, the better the molding properties of the powder. SH-3A can be seen to

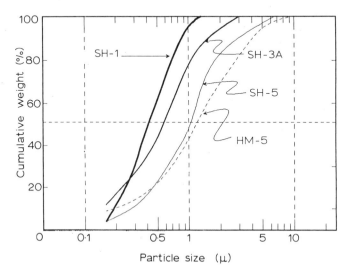

Fig. 2.4. Particle size distribution of four representative powders.

Table 2.3
Molding Properties of Four Representative Powders (green density—cold isostatically pressed at 1 ton/cm² after mixing for 20 h with 6% Y_2O_3, 4% Al_2O_3, and ethanol)

Sample	Green density (g/cm³)	Oil absorption (ml/100 g)
SH-1	1·83	25
SH-3A	1·88	20
SH-5	1·78	27
HM-5	1·79	28

have extremely good molding properties. The sintering properties of the four pressed powders are shown in Fig. 2.5. SH-1, which has the finest particle size, and SH-3A, which offers the best molding properties, are easy to sinter, even at low temperatures. Sintering was performed for 2 h under a nitrogen partial pressure of 9 kgf/cm². Onoda Cement manufactures and markets these and other powders with similar characteristics.

2.5 SUMMARY

At Onoda Cement powders are manufactured using strict quality control standards. Since sintering aids and mixing and molding techniques differ from customer to customer, however, specifications suited to these processes are required. Onoda Cement plans

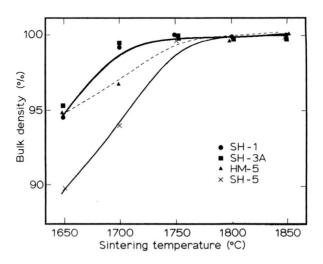

Fig. 2.5. Sintering properties of four representative powders.

to continue the supply of low-cost silicon nitride raw material powders with few irregularities. As mentioned earlier, Onoda Cement is also researching the characteristics of further powders, as yet unannounced, and developing powders suited to a wide variety of different processes, and will be providing these to its customers.

REFERENCES

1. Riley, F. L., Nitridation and reaction bonding. *Nitrogen Ceramics*, ed. F. L. Riley, p. 265.
2. Japanese Patent No. 54-15500, Kokai Tokkyo Koho, 1979.
3. Japanese Patent No. 54-24300, Kokai Tokkyo Koho, 1979.
4. Japanese Patent No. 54-120298, Kokai Tokkyo Koho, 1979.
5. Mitomo, M., *Refractories*, **38** (1986), 561.
6. Kanzaki, S., *Function and Materials*, **6** (1986), 5.
7. Kanazawa, T., *Solid Handling Proc. Ind.*, **18** (1986) 70.

3 | NKK's Silicon Nitride Powders: SIN Series

K. NAKAGAWA & M. KATO

ABSTRACT

NKK's Si_3N_4 powders are produced by nitridation of metallic silicon, ensuring a supply of high-quality, economical powder. In contrast to the high green densities and slurry concentrations of powders with an average grain size of 1·2 μm, 0·7-μm powder is liable to give lower densities and slurry concentrations. This drawback has been overcome by a special treatment resulting in the improvement of sintering characteristics. Flexural strengths of specimens sintered with the addition of 6% Y_2O_3 + 2% Al_2O_3 are over 70 kgf/mm² for normal pressure sintering, over 80 kgf/mm² for gas pressure sintering, and over 90 kgf/mm² for hot-pressed or HIPped samples.

3.1 INTRODUCTION

Silicon nitride ceramics are well balanced materials with high strengths (at both normal and high temperatures), a high fracture toughness, and a high thermal shock resistance. In recent years, research for Si_3N_4 has been particularly active in the field of high-temperature, high-strength applications.

Sizeable future increases in the applications for silicon nitride ceramics will require the resolution of several problems. One of these involves raw material powders: the present workers believe that improved quality and reduced cost are necessary.

NKK began basic research concerning silicon nitride powders

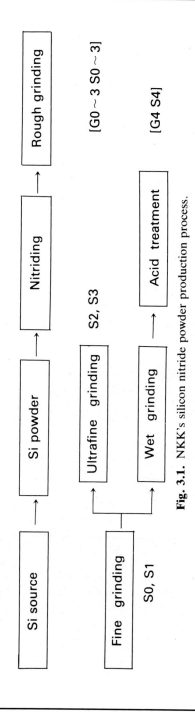

Fig. 3.1. NKK's silicon nitride powder production process.

in 1981 and built a series of production facilities in 1984. The firm is currently engaged in the production and marketing of silicon nitride powders. This chapter presents a brief summary and characterization of NKK's SIN silicon nitride powders.

3.2 PRODUCTION PROCESS

At present, most silicon nitride powders are produced using one of the following methods; direct nitriding of metallic silicon, silica reduction, and imide thermal decomposition. At NKK, the first method is used because it offers stable, economical production and high green densities. The direct nitriding process can be broken down into three main steps; raw material processing, nitriding, and grinding—an outline is given in Fig. 3.1.

The raw material metallic silicon to be used is selected from the available range of general industrial-grade silicon to single-crystal silicon, according to the chemical composition desired of the end product.

Fig. 3.2. External view of nitriding furnace.

In this method, the nitriding step is particularly important. At NKK the optimal conditions were determined as the result of various investigations, and improved productivity and stable product quality have been achieved by employing a unique temperature control method in a ~60 m^3 nitriding furnace (see Fig. 3.2) and a forced cooling system.

A large-scale mill with silicon nitride balls is employed in grinding to prevent the introduction of impurities. In addition, several-step air classification is performed according to particle size standards to stabilize product particle size.

3.3 POWDER CHARACTERISTICS

Product standards are composed of three characteristics: chemical composition (G grade); particle size (S grade); and crystal composition (α-type silicon nitride content, A grade). These are shown in Table 3.1. Powders from G0S0 to G3S3 are produced using dry grinding, while for G4S4 wet grinding is introduced.

Table 3.1 Varieties and Quality of Silicon Nitride Powders

Symbol	Chemical composition (G grade) (wt%)				
	Si	N	Fe	Al	Ca
G-0			<1·0	<0·7	<0·5
G-1			<0·8	<0·5	<0·3
G-2	57–59·5	37·5–39·0	<0·5	<0·5	<0·2
G-3			<0·2	<0·3	<0·1
G-4			<0·1	<0·2	<0·1

Symbol	Particle size (S grade) (µm)		Crystal composition (A grade) (wt%)	
	Maximum particle size	D_{50}	Symbol	α-Si_3N_4
S-0	<74	—	A-0	<50
S-1	<44	4·0	A-6	60–70
S-2	<10	1·2	A-7	70–80
S-3	<6	0·7	A-8	80–90
S-4	<4	0·7	A-9	>90

Table 3.2
Typical Characteristics of NKK's SIN Silicon Nitride Powders

Grade	Chemical composition					Average particle size (μm)[a]	Specific surface area (m²/g)[b]	Density characteristics		
	N	Fe	Al	Ca	O			Bulk	Uniaxial pressed[c]	CIPped[d]
G3S2	38·7	0·16	0·15	0·03	1·5	1·2	6	0·42	1·74	2·01
G3S3	38·4	0·16	0·15	0·03	1·8	0·7	9	0·36	1·45	1·72
Improved G3S3	38·0	0·16	0·15	0·03	2·2	0·7	11[e]	0·76	1·65	1·95
G4S4	39·0	0·03	0·08	0·03	1·1	0·7	12	0·44	1·69	1·97

[a] Microtrack SPA; [b] BET; [c] 400 kgf/cm²; [d] 2000 kgf/cm²; [e] 10–13 m²/g available.

NKK also produces non-standard powders having a variety of characteristics based on customer-provided specifications.

Typical values for average particle size, specific surface area, and density characteristics, all of which are considered particularly important as powder characteristics, are shown in Table 3.2. Scanning electron micrographs (SEM) of various powders are shown in Fig. 3.3.

3.4 SLURRY, MOLDING, AND SINTERING CHARACTERISTICS

In addition to meeting certain standards for sintered body characteristics, silicon nitride powders must offer good slurry characteristics, molding properties, and sintering properties for use in industrial production. One example of the results of tests carried out at NKK is described below.

3.4.1 Slurry Characteristics

The powder is often slurried when sintering aids are added. Stable slurry with high concentration and low viscosity is generally preferable. Numerous aspects of the relationship between these characteristics and powder characteristics remain unclear. Here, the results of investigations into the relationship between viscosity and concentration in the preparation of an aqueous system slurry will be discussed. A 1-liter pot mill with Si_3N_4 balls is used to obtain slurry of a powder and pure water with a specified concentration, and the relation between dispersion time and viscosity is measured. Viscosity is measured using a Brookfield-type viscometer. Results are shown in Fig. 3.4.

In Fig. 3.4, measurements for G3S3 were taken at a slurry concentration of 50%, while those for G3S2 and the improved G3S3 (three types are available, with BET specific surface areas of 9, 11, and 13 m^2/g) were taken at a 70% concentration. As can be seen from the graph, it is difficult to increase the concentration of the 0·7-μm (average particle size) S3 powder, but

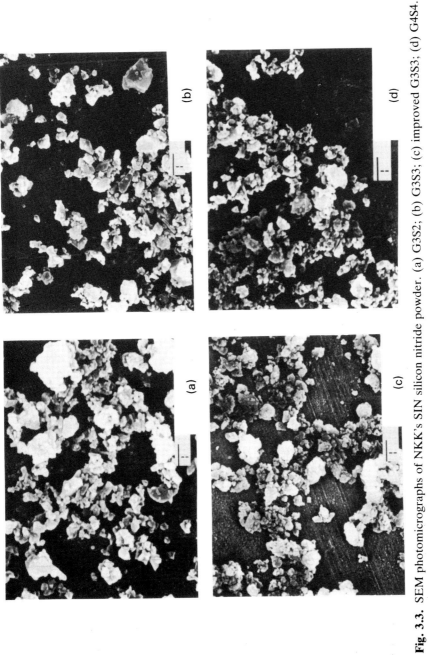

Fig. 3.3. SEM photomicrographs of NKK's SIN silicon nitride powder. (a) G3S2; (b) G3S3; (c) improved G3S3; (d) G4S4.

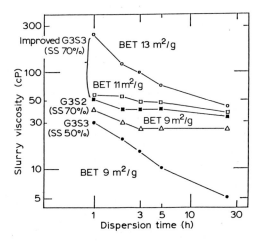

Fig. 3.4. Slurry characteristics of NKK's SIN silicon nitride powder.

improved S3, with an identical particle size, offers significantly improved slurry characteristics. The 1·2-μm S2 powder also shows good slurry characteristics.

3.4.2 Molding and Sintering properties

In order to evaluate the effect of a powder on the molding and sintering processes, various sintering methods were used on testpieces. The results are described below.

3.4.2.1 Normal-pressure Sintering

A sintering aid of 6% Y_2O_3 + 2% Al_2O_3 and dispersants were added to the aqueous system, and dispersal mixing was performed for 1 h in an Attritor. PVA and waxes were also added to prepare a granule by spray drying. This powder was uniaxially pressed at 400 kgf/cm^2 and isostatically cold-pressed at 2 ton/cm^2. It was then dewaxed and sintered at normal pressure for 2 h at 1750°C. A 60-mm diameter circular die was used for pressure forming. The relation between post-sintering weight and bulk density for each testpiece is shown in Fig. 3.5. A correlation between the sample weight and bulk density shown in this figure may have been affected by short-time sintering for 2 h at 1750°C.

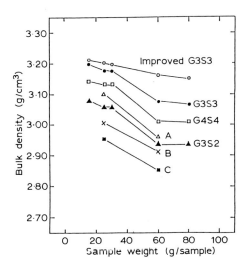

Fig. 3.5. Results of normal-pressure sintering at 1750°C for 2 h (6% Y_2O_3 + 2% Al_2O_3; attrition-milled for 1 h).

Improved G3S3, with a fine particle size and a high specific surface area, yielded the highest bulk density. It was followed by G3S3, G4S4, and G3S2 respectively.

A cross-sectional SEM photomicrograph of improved G3S3A7 is shown in Fig. 3.6(a). Despite having an α-content of about 75%, this powder was observed to have relatively many needle grains and a maximum particle size of 2–3 μm.

3.4.2.2 Gas-pressure Sintering

Using the same 6% Y_2O_3 + 2% Al_2O_3 system, a powder was mixed in hexane for 20 h using a ball mill. After drying and grading, the mixture was uniaxially pressed and CIPped at 3 ton/cm^2 and gas-pressure sintered at 1750°C for 10 h under an N_2 pressure of 10 atm. The results are shown in Table 3.3.

Cross-sectional SEM photomicrographs for sintered bodies obtained from G3S3A7 and G3S3A9 are shown in Figs 3.6(b) and (c). With A7 (α-content ~ 70%), sintering at 1750°C for 10 h showed both grain growth and a reduction in needle grains. A9 (with a 92% α-content) showed an abundance of needle grains and a reduction of grain growth. This has a major effect on the strength of the sintered body.

Fig. 3.6. Cross-sectional SEM photomicrographs of various sintered bodies: (a) improved G3S3A7, normal-pressure sintered; (b) G3S3A7, gas-pressure sintered; (c) G3S3A9, gas-pressure sintered; (d) G3S3A7, hot-pressed; (e) improved G3S3A7, HIPped.

Table 3.3
Results of Gas-pressure Sintering for NKK's SIN Powder

Grade	CIP density (g/cm^3)	Bulk density (sintered) (g/cm^3)	Weight loss in sintering (wt%)	Flexural strength at R.T. (kgf/mm^2)	Flexural strength at elevated temperature of 1200°C (kgf/mm^2)
G3S2A8	2·05	3·20	1·8	62	41
G3S3A7	1·92	3·24	2·5	60	35
G3S3A8	1·90	3·24	2·6	75	53
G3S3A9	1·88	3·24	2·5	79	50
Improved G3S3A8	2·02	3·24	2·8	83	46

6% Y_2O_3 + 2% Al_2O_3; 20 h ball-mill mixing; CIPped at 3 ton/cm²; sintered at 1750°C for 10 h under 10-atm N_2 pressure.

Table 3.4
Results of Special Sintering of NKK's SIN Powder

Sintering method	SIN grade	Sintering aid (wt%)	Bulk density (sintered) (g/cm^3)	Flexural strength at R.T. (kgf/mm^2)	Sintering conditions
HP	G3S3A7	$6Y_2O_3 + 2Al_2O_3$	3·24	95	1800°C for 2 h at 200 kgf/cm²
Normal-pressure sintering	Improved G3S3A7	$6Y_2O_3 + 2Al_2O_3$	3·20	76	1750°C for 2 h
Normal-pressure sintering + HIP	Improved G3S3A7	$6Y_2O_3 + 2Al_2O_3$	3·24	93	1750°C for 2 h at 1900 atm

3.4.2.3 Special Sintering

Hot-pressed and isostatically hot-pressed sintering were also conducted for the 6% Y_2O_3 + 2% Al_2O_3 system. In the case of hot-pressing, 30 g of the G3S3 granulated powder was set in a 40-mm diameter carbon die coated with BN powder; sintering was performed at 1800°C for 2 h under a pressure of 200 kgf/cm². For isostatic hot-pressing, a sintered body with a density of 3·20 g/cm³, obtained from improved G3S3A7, was pressed at 1900 atm, 1750°C for 3 h. The results of these tests are shown in Table 3.4. Changes in the density increase (calculated from the displacement measured during hot-pressing) as a function of time are shown in Fig. 3.7. These measurements show that rearrangement of grains and contraction in sintering begin to occur in the temperature range 1000–1200°C, with sintering mainly occurring around 1600°C and being virtually completed approximately 30 min after reaching 1800°C. Cross-sectional SEM photomicrographs of hot-pressed and HIPped sintered bodies are shown in Figs 3.6(d) and (e). In both cases densification is virtually complete and no grain growth has occurred.

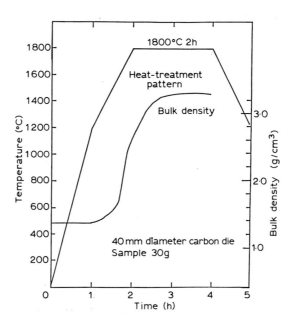

Fig. 3.7. Density change in hot-pressing (G3S3A7).

3.5 SUMMARY

The powder characteristics of NKK's SIN and the characteristics of its sintered bodies obtained in various sintering methods have been discussed. Numerous methods are available for the fabrication of silicon nitride ceramics (differing in sintering aids, molding methods, sintering methods, etc.) and the requirements for raw material powders are correspondingly diverse. As a producer of powders, NKK hopes to deepen ties with sintered-product manufacturers, strengthening the relationship in the future to expand the range of applications for silicon nitride powders.

NKK will continue to supply stable-quality, economical raw material powders, with characteristics suited to customers' requirements.

4 | Characterization and Synthetic Process of Si_3N_4 Material Powders

M. NAKAMURA, Y. KURANARI & Y. IMAMURA

ABSTRACT

Silicon nitride, one of the well known advanced ceramics, is expected to be of use in future materials for industry because of its many excellent properties. Si_3N_4 material powders are produced mainly by three methods: silica reduction, silicon metal nitridation, and silicon chloride and ammonia reaction.

In this chapter two processes are described; silicon metal nitridation, and silicon chloride and ammonia reaction. The former consists of the purification of MG–Si, metallic silicon nitridation, crushing and milling of Si_3N_4 lumps, acidizing, and drying. The latter consists of synthesizing a Si_3N_4-precursor with a gaseous reaction of $SiCl_4$ and NH_3, sublimation of NH_4Cl, calcining (crystallization), and deflocculation. Manufactured Si_3N_4 powders are analyzed and tested.

The obtained results are summarized as follows:

(1) The impurities contained in MG–Si (such as Fe, Al, and Ca) to the order of 1000 ppm are concentrated in the silicon grain boundaries by fusion–solidification for easy purification. After acidizing, these impurities decrease to around 100 ppm.

(2) The qualities of Si_3N_4 powder manufactured using the

former method are excellent. A typical analysis shows Fe, Al, and Ca contents of 29, 54, and 23 ppm, respectively, with an oxygen content of 0·86%. *Furthermore, the specific surface area is* 22 m^2/g, *and the mean particle diameter is* 0·5 µm.

(3) *When the reaction temperature is low, the $NH_3/SiCl_4$ ratio increases together with the Si_3N_4-precursor yield rate, but the crystallized Si_3N_4 yield rate decreases. Hence, the preferable $NH_3:SiCl_4$ ratio is concluded to be about 6.*

(4) *The qualities of Si_3N_4 powder manufactured using silicon chloride and ammonia reaction are also very good; contained Fe, Al, and Ca are 16, 18, and 2 ppm, respectively, and particle distribution is very uniform.*

4.1 INTRODUCTION

Silicon nitride is stronger than conventional oxide ceramics and possesses excellent resistance to heat, corrosion, and wear. However, since it is basically a brittle material it is quite vulnerable to minute surface scratches and internal defects, and major technological developments will be required to improve reliability. As part of this development, synthesis methods are being developed to produce high-purity, fine-grained raw material powders, techniques for evaluating synthesized powder characteristics are being studied, and work is proceeding on the understanding of the relationship with molding and sintering properties. This chapter touches especially on methods of synthesizing high-quality silicon nitride powders, and the characteristics thereof.

4.2 SILICON NITRIDE RAW MATERIAL POWDERS

Silicon nitride (Si_3N_4) is an artificial mineral with strong covalent bonding properties and two crystal types, α and β, as shown in Table 4.1. The former type is commonly referred to as the low-temperature type. Although the majority of silicon nitride

Characterization and Synthetic Process

Table 4.1 Physical Properties of Si_3N_4

Crystal system	α, trigonal; β, hexagonal
Lattice constant	
α	$a_0 = 7\cdot7520 \pm 0\cdot0007$ Å
	$c_0 = 5\cdot6198 \pm 0\cdot005$ Å
β	$a_0 = 7\cdot608 \pm 0\cdot005$ Å
	$c_0 = 2\cdot911 \pm 0\cdot001$ Å
True density	
α	$3\cdot168$ g/cm^3
β	$3\cdot192$ g/cm^3
Colour	Grey to thin yellowish-brown
Sublimation point	1870°C
Specific heat	$0\cdot26$ cal/g/°C

compacts are β-type, raw material powders with a high α-content are preferred during sintering because of the ease with which columnar crystals develop, etc.

Silicon nitride raw material powder characteristics can be broadly grouped into three categories: (1) those concerning particle size and grain morphology; (2) those concerning chemical composition and impurities; and (3) those concerning crystallinity (see Fig. 4.1). All of these are closely related to the starting material and the synthesis method and conditions. Since some items, such as oxygen content and secondary flock, are also affected by storage methods and handling, particular care must be paid in the handling of high-purity, ultrafine powders.

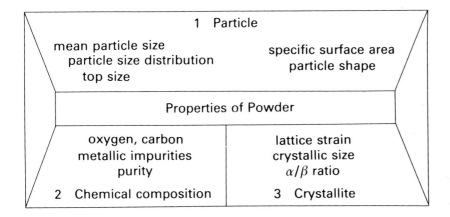

Fig. 4.1. Characterization of Si_3N_4 powder.

Table 4.2 Synthetic Process of Si$_3$N$_4$ Powder

Synthetic process of Si$_3$N$_4$ powder		Reaction
Silica reduction method		$3SiO_2 + 6C + 2N_2 = Si_3N_4 + 6CO$
Silicon metal nitridation method		$3Si + 2N_2 = Si_3N_4$
Silicon chloride and ammonia method	Gas	$3SiCl_4 + 4NH_3 = Si_3N_4 + 12HCl$ (amorphous) $Si_3N_4(\text{amorphous}) = Si_3N_4(c)$
	Liquid	$3SiCl_4 + 16NH_3 = 3Si(NH)_2 + 12NH_4Cl$ $3Si(NH)_2 = Si_3N_4 + N_2 + 3H_2$

The main methods of synthesizing silicon nitride raw material powders are shown in Table 4.2. In the silica reduction method, high-purity, ultrafine SiO$_2$ and C are heated in an N$_2$ atmosphere, reduction and nitriding are simultaneously performed, and excess C is removed by after-treatment. Although it is possible to obtain a powder with a high α-content and a narrow particle size distribution, it is said to be difficult to simultaneously reduce both the SiO$_2$ and the C contained in the silicon nitride powder.

Table 4.3 Commercial Si$_3$N$_4$ Powders

	Si$_3$N$_4$ powder				
	M-1	M-2	M-3	M-4	M-5
Specific surface area (m^2/g)	12·1	21·6	15·6	13·4	8·7
Mean particle diameter (μm)	0·65	1·36	1·71	0·81	0·90
Phase content (%)	91·4	95·8	97·1	86·8	98·5
Total oxygen (%)	0·89	2·44	2·53	2·22	2·00
Total carbon (%)	0·24	0·16	0·23	0·18	0·90
Metallic impurities (ppm)					
Fe	280	100	40	30	70
Al	1230	380	30	10	2000
Ca	1030	360	10	5	100

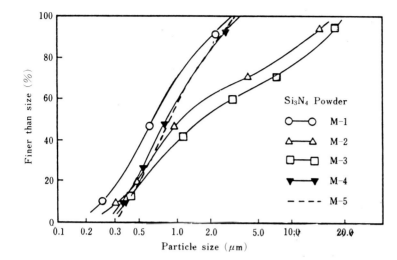

Fig. 4.2. Particle size distribution of commercial Si_3N_4 powders.

In the metallic silicon process, Si powder is heated and nitrided in an N_2 atmosphere. While the reaction is relatively simple, control of the α-content is difficult, and there is a tendency for impurities to be introduced during grinding, as well as being contained in the raw metallic silicon. In the silicon chloride and ammonia method, Si-halide and NH_3 are reacted in the vapor or liquid phase, silicon nitride precursors (nitrogen-containing silanized compounds–amorphous silicon nitride) are synthesized, and the precursors are then crystallized by heating. Although a high-purity product can be obtained, needle crystals are often formed, and halogen-induced corrosion and by-products (NH_4Cl, HCl) must be removed.

Characteristics of, and particle size distributions for, some of the higher-quality commercial silicon nitride raw material powders available today are shown in Table 4.3 and Fig. 4.2. Although powder characteristics considered favorable for fine ceramics raw materials vary, depending on molding and sintering methods, in general five characteristics are sought after: (1) fine grain, (2) anoxia, (3) high purity, (4) equiaxed grain morphology, and (5) phase control. Work also continues on the improvement and stabilization of product quality.

4.3 SYNTHESIS OF SILICON NITRIDE RAW MATERIAL POWDERS USING THE METALLIC SILICON METHOD

The basic process for the fabrication of high-quality silicon nitride powder using the metallic silicon method is shown in Fig. 4.3. Methods of refining industrial silicon in particular, as well as increasing the scale of nitriding operations, are discussed in this section.

4.3.1 Refining Industrial Silicon

The main metallic impurities contained in industrial Si are Fe, Al, Ca, etc.; ordinarily, anything from 100 to several thousand ppm of each are present. If the Si is nitrided in this state, the

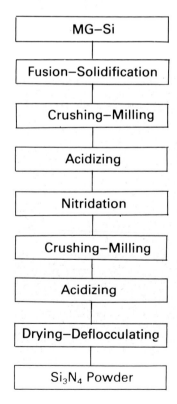

Fig. 4.3. Fundamental process of silicon metal nitridation.

(a)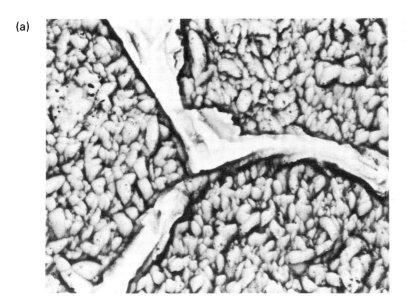

(b)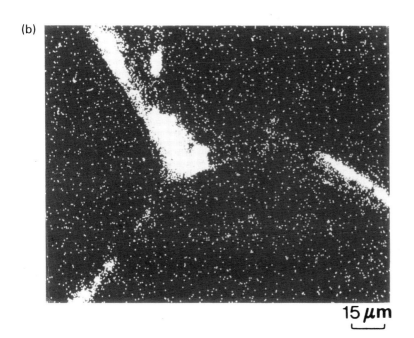

Fig. 4.4. Iron distribution in MG–Si: (a) SEM photomicrograph; (b) EPMA image.

later removal of these metallic impurities through after-treatment is extremely difficult. At first, the Fe, Al, and Ca contents of the metallic silicon were successfully reduced to less than several tens of ppm through unidirectional solidification. However, since this method requires a large temperature gradient for the solidification surface, the diameter of the solidification ingots is limited, and the metallic impurity concentration portion accounts for 10–30% of the material, making scale increases in this state unfavorable. Rather than concentrating the metal impurities in one direction, the present workers have developed the fusion–solidification–acidization method, in which the impurities are microscopically concentrated (gathered along the grain boundaries) for refining. As shown in Fig. 4.4, by concentrating the metallic impurities (e.g., Fe) along the Si grain

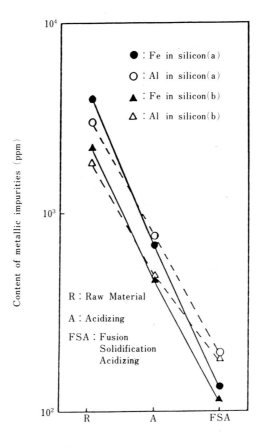

Fig. 4.5. Dependence of metal impurities on treating with acid of MG–Si.

boundaries and performing fusion–solidification of the metallic silicon, the size of the Si crystal grain is increased, and by removing the metallic impurities concentrated along the grain boundaries with later grinding and acidizing, far greater reductions in metallic impurities are possible than when acidization alone is used (see Fig. 4.5).

4.3.2 Increasing the Scale of Nitriding Operations

The nitriding reaction of metallic silicon is accomplished using a large amount of heat and often develops into a so-called 'runaway reaction,' making the fabrication of silicon nitride with a high α-content difficult. Therefore, in order to develop larger-scale nitriding technologies, the reaction must somehow be slowed down. The present workers have conducted tests concerning nitriding patterns of the type shown in Fig. 4.6, etc., and bench-scale nitriding technology has been virtually established.

The nitride ingots thus obtained are processed as shown in Fig. 4.3 to produce a silicon nitride powder. The characteristics of a high-quality silicon nitride powder obtained using metallic silicon are shown in Table 4.4.

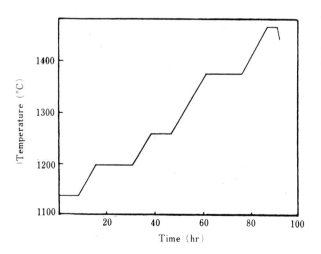

Fig. 4.6. Nitridation pattern of silicon metal.

Table 4.4
Properties of the Synthesized Si_3N_4 Powder (SN–GX)

True density (g/cm^3)	3·12
Specific surface area (m^2/g)	22·0
Mean particle diameter (μm)	0·5
α-phase content (%)	92·1
Metallic Impurities (ppm)	178
Fe	29
Ca	23
Al	54
B	1
Ba	2
Co	30
Cr	4
Cu	4
Mg	8
Mn	2
Mo	8
Na	2
Ni	1
Sr	<1
Ti	<1
V	1
W	4
Zn	<1
Zr	<1
K	<1
Purity (%)	98·8
Total oxygen (%)	0·86
Total silicon (%)	58·9
Total nitrogen (%)	38·8
Total carbon (%)	0·26
Free silicon (%)	0·02
Free silica (%)	0·12

4.4 SYNTHESIS OF SILICON NITRIDE RAW MATERIAL POWDERS USING THE SILICON CHLORIDE AND AMMONIA METHOD

The basic process involved in the silicon chloride and ammonia method is shown in Fig. 4.7. In the first half of the process, silicon nitride precursors are obtained using the low-temperature vapor phase method, while large amounts of ammonium chloride are simultaneously produced as a by-product. In the latter half of

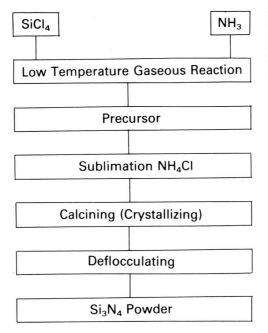

Fig. 4.7. Fundamental process of silicon chloride and ammonia reaction.

the process, silicon nitride powder is fabricated using sublimated removal of ammonium chloride and heat-induced crystallization.

4.4.1 Synthesis of Precursors

Cost-effective $SiCl_4$ is used as silicon halide, and intermediates (silicon nitride precursors + by-product NH_4Cl; precursors with a N:Si molar ratio of ~2) are synthesized using a vapor-phase reaction with NH_3 gas. The intermediate yield varies depending upon the $NH_3:SiCl_4$ ratio of the raw material gas, synthesis temperature, etc.; the lower the synthesis temperature, the greater the domination of the second of the following two reactions (see Fig. 4.8):

$$3SiCl_4 + 4NH_3 = Si_3N_4 \text{ (amorphous)} + 12HCl \quad (1)$$

$$3SiCl_4 + 18NH_3 = 3Si(NH)_2 + 12NH_4Cl \quad (2)$$

As a result, when the reaction rate for $SiCl_4$ intermediate synthesis is taken into consideration, a high ratio of NH_3 to $SiCl_4$

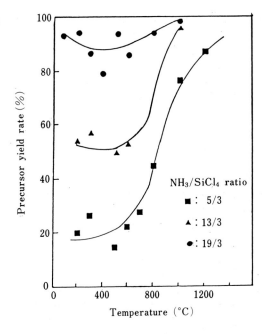

Fig. 4.8. Relation between the precursor yield rate and the reaction temperature.

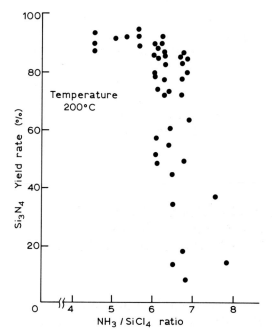

Fig. 4.9. Relation between the condition of the Si_3N_4 yield rate and the $NH_3:SiCl_4$ ratio.

is favorable at low synthesis temperatures, but since the Si yield drops during crystallization of the obtained intermediates, an optimum $NH_3:SiCl_4$ ratio does exist (Fig. 4.9). The reasons why yield drops during crystallization are currently under investigation, but it has been confirmed that the composition of the intermediates is slightly affected by the ratio of NH_3 to $SiCl_4$.

4.4.2 Removal of NH_4Cl and Heat-induced Crystallization

It is possible to remove the NH_4Cl contained in sizeable quantities in the intermediates by heating, but the resulting device corrosion becomes a problem. As a result, further investigation of the materials is needed.

Crystalline silicon nitride powder can be obtained by heating intermediates (precursors) from which the NH_4Cl has been removed above 1400°C in an atmosphere containing N_2. Although the intermediates (the starting material) ordinarily contain 1% oxygen, virtually all of this is released from the system during heat-induced crystallization, and the oxygen content of the silicon nitride powder is approximately 1%.

Heating conditions have an extremely large influence on crystallization: the higher the temperature and the longer it is maintained, the larger the silicon nitride powder particle size (i.e., the smaller the specific surface area) and the more likely the formation of needle crystals. The influence of the atmosphere itself appears to be relatively slight. Characteristics of a high-quality silicon nitride powder prepared using this method are shown in Table 4.5.

4.5 CHARACTERISTIC COMPARISON OF POWDERS OBTAINED FROM METALLIC SILICON AND SILICON CHLORIDE AMMONIUM METHODS

The silicon chloride ammonium method is generally held to offer higher purity, but if silicon is sufficiently refined and the

Table 4.5 Properties of the Synthesized Si_3N_4 Powder (Silicon Chloride Ammonia Reaction)

True density (g/cm^3)	3·13
Specific surface area (m^2/g)	8·6
Mean particle diameter (μm)	0·9
α-phase content (%)	98·0
Metallic Impurities (ppm)	166
Fe	16
Ca	2
Al	18
B	<1
Ba	<1
Co	16
Cr	4
Cu	<1
Mg	2
Mn	<1
Mo	<1
Na	<2
Ni	88
Sr	<1
Ti	1
V	<1
W	<1
Zn	7
Zr	<1
K	<1
Purity (%)	99·0
Total oxygen (%)	0·65
Total silicon (%)	59·5
Total nitrogen (%)	38·8
Total carbon (%)	0·13
Free silicon (%)	—
Free silica (%)	0·35

introduction of impurities strictly monitored, a combined Fe + Ca + Al ratio of approximately 100 ppm is possible with the metallic silicon method.

Grain morphology differs significantly. The high-quality silicon nitride powder prepared with the metallic silicon method and shown in Fig. 4.10 falls in the sub-micron range and appears to have relatively uniform particle size, but it has a slightly wider particle size distribution than the high-quality silicon nitride powder prepared with the silicon chloride ammonium method and shown in Fig. 4.11. The former powder also has an angular

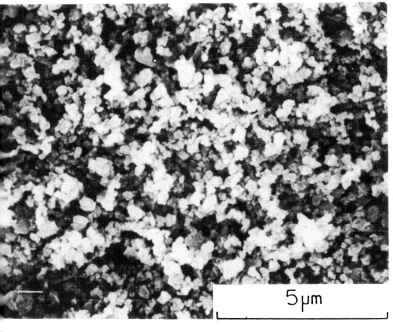

Fig. 4.10. SEM photomicrograph of the Si_3N_4 powder synthesized by silicon metal nitridation.

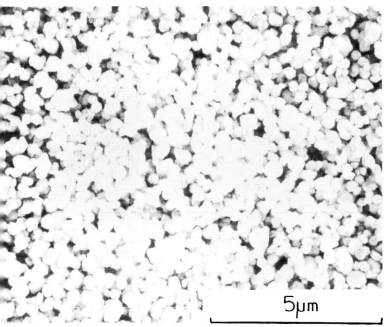

Fig. 4.11. SEM photomicrograph of the Si_3N_4 powder synthesized by silicon chloride and ammonia reaction.

Fig. 4.12. Particle size distribution of Si_3N_4 powders.

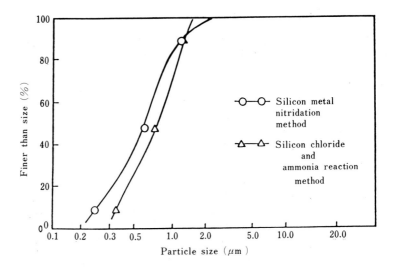

grain morphology and contains 0·1–0·2-μm ultrafine particles. This is also shown clearly in particle size distributions of the high-quality silicon nitride powders produced by either method and shown in Fig. 4.12. However, when molding properties are taken into consideration, powders with relatively wide distributions are favored, and the same results have been obtained with the measurement of tap density, etc.

Turbulence has been predicted in the crystal structure of powders obtained using the metallic silicon method, due to the grinding and acidizing processes. Therefore, crystallite size and lattice strain (see Table 4.6) were calculated from the equation $\beta \cos \theta / \lambda = 0·89/L + 2\varepsilon \sin \theta / \lambda$ where β is the half-value width of each diffraction line, L is the average crystallite size, and ε is the size of the internal crystal strain (lattice strain) distribution, assuming a Cauchy distribution.

Table 4.6 Crystallinity of Si_3N_4 Powders

	Silicon nitridation	*Si chloride +ammonia*	*M-1*	*M-2*	*M-3*	*M-4*	*M-5*
Crystallite size (Å)	580	650	720	570	550	550	590
Lattice strain ×10⁻³	2·40	1·90	2·70	1·90	2·00	2·50	3·00

As a result, the metallic silicon was found to exhibit greater lattice strain and smaller crystallites. However, when commercial silicon nitride powders are included in the comparison, the correlation between the grinding history and the synthesis method disappears.

Figure 4.13 is a transmission electron micrograph of the powder synthesized using the metallic silicon method. Although turbulence in 2–4 layers of crystals (surface SiO_2 layers) can be seen along the periphery, there is no internal turbulence.

For reference, Fig. 4.14 shows the results of an experiment in which MG–Si was nitrided without being refined, and the relationship between the degree of grinding and oxygen content was investigated. It is assumed that the oxygen content increases both on the surface and internally together with grinding, and when an estimate is made from the specific surface area, the total oxygen content of products fabricated using the metallic silicon method is expected to be quite large. However, oxidation prevention measures have allowed the production of a powder with an oxygen content of less than 1%.

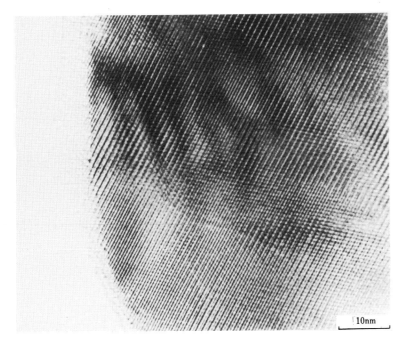

Fig. 4.13. TEM photomicrograph of the Si_3N_4 powder synthesized by silicon metal nitridation.

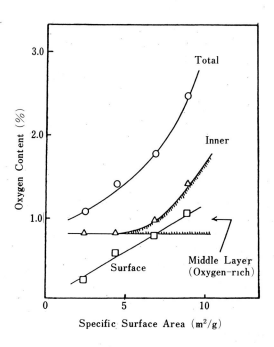

Fig. 4.14. Relation between the surface oxygen content of Si_3N_4 powder and the specific surface area of dry-ground Si_3N_4 powder.

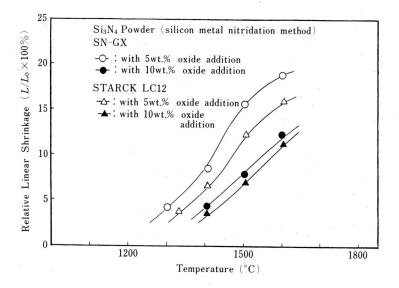

Fig. 4.15. Dependence of relative linear shrinkage on sintering temperature.

Although it is possible that sintering properties vary considerably, depending on the type of tests conducted, when compared with Starck LC 12 (see Fig. 4.15), which has roughly the same specific surface area, powders prepared with the metallic silicon method were judged to offer a higher ease of sintering.

4.6 SUMMARY

Together with the development of synthesis technology (increases in production scale, etc.) for silicon nitride raw material powders, the characterization of these powders is very important. In future, the authors plan to continue research on product quality and uniformity while feeding back the results of molding, sintering, and powder evaluations.

ACKNOWLEDGMENTS

The current work is just one part of the research and development being conducted on fine ceramics as Basic Technologies for Future Industries. The authors are deeply grateful to the Agency of Industrial Science and Technology of the Ministry of International Trade and Industry (MITI) for permission to publish this contribution.

REFERENCES

1. Wild, S., Grieveson, P. & Jack, K. H., The crystal structure of α and β silicon and germanium nitrides. In *Special Ceramics*, Vol. 6, ed. P. Pepper. British Ceramics Society, 1972, pp. 386–95.
2. Kijima, K., Tanaka, H. & Setaka, N. Preparation of α-Si_3N_4 with low oxygen impurity. *Yogyo-Kyokai-Shi,* **84** (1976), 14.
3. Colquhoun, I., Wild, S., Grieveson, P. & Jack, K. H., The determination of surface silica and its effect on the hot-pressing behavior of alpha-silicon nitride powder. *Proc. Br. Ceram. Soc.,* **22** (1973), 207.

4. MITI, Technology assessment of fine ceramics. Report 1980.
5. Tsuge, A., Process technique of ceramic powder available sintering. Paper presented at 11th Heat Resistance Material Seminar, Tokyo, February, 1980.
6. Lange, F. F., Phase relations in the system Si_3N_4–SiO_2–MgO and their interrelation with strength and oxidation. *J. Am. Ceram. Soc.*, **61** (1978), 53–6.
7. Synthetic materials for engineering ceramics. *Ceramics Data Book*, Kogyo Seihin Gijutu Kyokai, 1985, pp. 101–6.
8. Inagaki, Effective debye parameter of polycrystalline materials. *J. Mat.*, **8** (1973), 312–16.

5 α-Si₃N₄ Powder Produced by Nitriding Silica Using Carbothermal Reduction

T. ISHII, A. SANO & I. IMAI

ABSTRACT

Silicon nitride is noteworthy as a high-strength, heat-resistant engineering material. The silicon nitride produced by nitriding silica using carbothermal reduction (silica reduction) is highly regarded, because sintered materials made from the powder have excellent strength at high temperatures. The reason for this remains unclear. The synthesizing process and typical properties of the silicon nitride powder produced by the silica reduction method are briefly described here.

5.1 INTRODUCTION

Silicon nitride is an attractive candidate as a high-temperature, high-strength material. In particular, α-Si₃N₄ powders produced using carbothermal reduction are highly valued because of the excellent strength at high temperatures that can be achieved in sintered bodies made from the powders. Although the relevant mechanism remains unclear, this chapter discusses the method of manufacturing α-Si₃N₄ powder using carbothermal reduction, and the characteristics of the powder thus obtained.

5.2 AN OUTLINE OF SILICA REDUCTION NITRIDING SYNTHESIS

5.2.1 Fabrication Process

In the synthesis of α-Si_3N_4 by carbothermal reduction, a fine raw material powder of silica, carbon and Si_3N_4 is blended and mixed at a specified ratio, and this raw material is then heated in flowing N_2 gas. After synthesis is completed, the material is decarbonized to remove excess carbon. Crushing or grinding is then performed as necessary to produce the final product. The flow chart of the manufacturing process is shown in Fig. 5.1. This chart will be followed in the following description of raw materials, mixing, synthesis, decarbonization, crushing/grinding, etc.

5.2.2 Raw Materials and Mixing

Formation of α-Si_3N_4 using carbothermal reduction is assumed to be due to the following elementary reactions:

$$SiO_2 + C \rightarrow SiO + CO \qquad (1)$$

$$SiO + C \rightarrow Si + CO \qquad (2)$$

$$3Si + 2N_2 \rightarrow Si_3N_4 \qquad (3)$$

The overall reaction is shown by the following equation:[1-3]

$$3SiO_2 + 6C + 2N_2 \rightarrow Si_3N_4 + 6CO \qquad (4)$$

The properties of the Si_3N_4 powder obtained according to eqn (4) are greatly affected by the raw material, the raw material blending ratio, and the synthesis conditions. The purity of the raw material used becomes the purity of the final product, and by using a high-purity raw material high-purity Si_3N_4 can be obtained.

Also important in obtaining high-quality Si_3N_4 powder is the blending ratio of carbon to silica. Although the theoretical weight ratio of SiO_2 to C is 1:0·4, in order to obtain uniform Si_3N_4 with a high nitride ratio it is necessary to make the carbon content larger than eqn (4) would suggest.

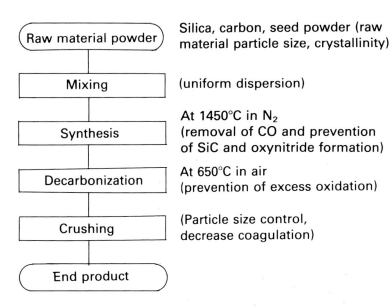

Fig. 5.1. Fabrication process for α-Si_3N_4 powder using carbothermal reduction. Items in parentheses show technical elements.

Seed crystals are also an important factor in controlling the particle size and grain morphology of the Si_3N_4 powder thus synthesized. Since nucleation alone does not produce sufficient Si_3N_4 for the reaction to proceed and results in non-uniform particle sizes, a specified amount of Si_3N_4 powder is added to the raw material in advance, as seed material. The Si_3N_4 formed according to eqn (4) accumulates around the particle of Si_3N_4 thus added as seed and induces crystal growth.

An experiment carried out to confirm this phenomenon[4] is shown in Fig. 5.2. Si_3N_4 whiskers were added at a weight ratio of 0·1–1·6 with respect to the SiO_2, and nitride synthesis was performed using carbothermal reduction. Whisker growth was then observed with scanning electron microscopy (SEM). Cases in which the weight ratio (with respect to SiO_2) of the whiskers added was 0·1, 0·15, 0·4, 0·8, and 1·6, respectively, are shown in Figs 5.2(a)–(e). The smaller the amount of whiskers (i.e., the smaller the surface area allowing accumulation), the greater the growth in the direction of the whisker diameters.

When only a small amount of whiskers was added, as in Figs 5.2(a) and (b), Si_3N_4 particles are present in relatively large amounts. This is believed to be due to nuclei formed by homogeneous nucleation that have grown into particles, or Si_3N_4

Fig. 5.2. SEM photomicrograph taken after whisker doping and synthesis. Weight of doped whiskers with respect to SiO_2 is (a) 0·1, (b) 0·15, (c) 0·4, (d) 0·8, (e) 1·6 and (f) whisker raw material.

particles contained in the raw material whiskers. As a result, close investigation is expected to be very difficult. However, if it is assumed that the Si_3N_4 formed is entirely deposited on the whisker surface and grows into crystals such that thickness increases uniformly at all locations, growth in the direction of the whisker length is ignored, and the correlation between growth in the direction of the diameter and the whisker additive amount is inferred, the following equation can be obtained:

$$d/d_0 = \sqrt{1 + 0·78/W} \qquad (5)$$

where d_0 is the diameter of the added whiskers, d is the diameter of the whiskers after growth, and W is the weight ratio of the whiskers and the SiO_2.

The relationship between d/d_0 and W is shown in Fig. 5.3. There is a correlation between the amount of added whiskers and growth in the diametral direction, and the actual measurements obtained from SEM photomicrographs closely coincide with values calculated from eqn (5). As a result, the type of Si_3N_4

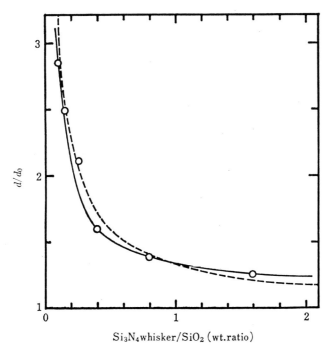

Fig. 5.3. Relation between whisker doping amount and post-synthesis whisker growth. —O—, Measured values; --O--, calculated values.

powder used for the seed crystal is extremely important, as it is no exaggeration to say that the properties of the seed powder become the properties of the synthesized powder and, in particular, the morphological properties thereof.

Various raw materials have been described. Equation (1) is a solid–solid reaction, and in order to make it occur efficiently and to maximize the uniformity of seed crystal growth uniform dispersion and mixing of the raw materials is necessary.

5.2.3 Synthesis

Synthesis conditions such as temperature, pressure, and N_2 gas flow are also important in determining the quality of the synthesized product and manufacturing costs. In general, the rate of a chemical reaction increases with temperature. In the case of the carbothermal nitriding reaction, however, eqn (1) is a solid–solid reaction, eqn (2) is a vapor–solid reaction, and eqn (3) is a vapor–vapor reaction. Since the factors determining the reaction rate are different in each case, it is impossible to generalize. Ordinarily, however, the speed of Si_3N_4 formation increases with temperature.

The carbothermal reduction is a reversible reaction. As can be seen from eqn (4), an increased CO partial pressure within the system will result in a reverse reaction, thereby hindering the nitriding reaction. In order to prevent this, CO gas must be promptly removed from the reaction system. In practice, the CO gas partial pressure can be kept at low levels by supplying more N_2 gas to the production furnace than is theoretically suggested.

Stable relationships between Si_3N_4, SiO_2, and SiC are shown in Fig. 5.4, with the reciprocal of the absolute temperature as the horizontal axis, the N_2 gas partial pressure as the vertical axis, and the CO gas partial pressure as a parameter.[5] Given an N_2 gas partial pressure of 1 atm, it can be seen that Si_3N_4 can exist as a stable phase at CO partial pressures of less than 0·345 atm and temperatures of less than 1435°C. Even when the CO partial pressure is less than 0·345 atm, SiC is stable at temperatures higher than 1435°C. From the standpoint of the reaction rate, a higher synthesis temperature is preferable, but in order to reduce the formation of SiC synthesis temperatures exceeding 1435°C are not favorable. It should also be noted that SiO_2 becomes

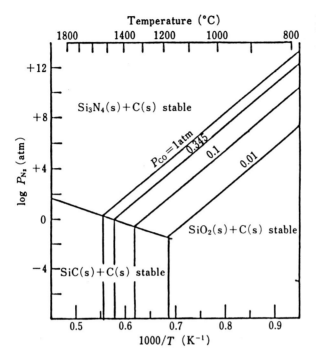

Fig. 5.4. Thermochemical correlation for stability of Si_3N_4, SiO_2, and SiC under coexistence with carbon.[5]

stable (and hence the nitriding reaction ceases to occur) when the CO gas partial pressure rises above 0·345 atm, even when the temperature is less than 1435°C.

5.2.4 Decarbonization and Crushing/Grinding

Carbothermal reduction synthesis is performed by mixing excess carbon with the raw material. As described above, decarbonization involves the removal, after nitriding synthesis, of excess (unreacted) carbon. In this process, C is reacted with O_2 and released from the reaction system as CO or CO_2 gas. In practice, this is done by heating the synthesized product to a temperature of about 650°C in air.

Crushing and grinding are performed as necessary to decrease the amount of coarse particles and coagulation in the synthesis powder, and thereby improve molding and sintering properties.

5.3 CHARACTERISTICS OF POWDERS SYNTHESIZED USING CARBOTHERMAL REDUCTION

General characteristics of powders synthesized using carbothermal reduction are as shown in Table 5.1. The advantages of this powder are a high α-phase content, fine grain, and a sharp particle size distribution, all of which result in a sintered body with high strength even at high temperatures. Figure 5.5 is an SEM photomicrograph of grade A-200 α-Si_3N_4. The particles are spherical, and particle size is relatively uniform.

The results of chemical analysis of A-200 are shown in Table 5.2, together with the particle size characteristics for the powder. Fe, Ca, Mg, and other metallic impurities are present at extremely low levels: also noteworthy are the relatively high contents of Al and C. Aluminum is introduced from the grinding media during the grinding process, and it can be applied effectively during sintering as a sintering aid.

When the sintering operation is carried out properly, C is believed to be the source of the excellent high-temperature strength possessed by sintered bodies. While it is technically possible to produce powders with even lower carbon contents, therefore, a certain amount is considered favorable.

The results of three-point bending strength tests conducted on sintered bodies made using A-200, at room temperature, 1000°C, and 1200°C, are shown in Fig. 5.6. The testpiece dimensions were $4 \times 3 \times 40$ mm and, for hot-pressed sintering, Y_2O_3 and Al_2O_3 were used as sintering aids. During normal-pressure sintering, Y_2O_3, Al_2O_3, and AlN were used as sintering aids. Hot-pressed sintered bodies had higher bending strengths than normal-pressure sintered bodies at all three temperatures, but the difference between the two was not as significant at 1200°C as

Table 5.1 General Properties of α-Si_3N_4 Powder Prepared Using Carbothermal Reduction

High α-phase content (>98%)
Uniform shape and small variation in particle size
Mean particle size <1 μm
Moderately high purity
Excellent sintered body characteristics

Fig. 5.5. SEM photomicrograph of A-200 α-Si_3N_4 powder.

Table 5.2 Characteristics of A-200 α-Si_3N_4 Powder (typical values)

Grade		A-200
Chemical composition (%)	Si	59·31
	N	37·5
	O	2·25
	C	0·69
	Free SiO_2	1·77
	Al	0·128
	Fe	0·007
	Ca	0·008
	Mg	0·002
Particle size characteristics (μm)	Mean particle size FSSS	0·67
	Microtrack	
	D10	0·40
	D50	0·85
	D90	2·17
	Tap density (g/cm^3)	0·89

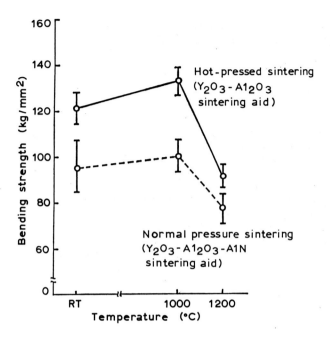

Fig. 5.6. Three-point strengths of A-200 sintered bodies at three temperatures.

at room temperature or 1000°C. Comparative examples are not shown in Fig. 5.6, but the strengths shown were higher than those possessed by sintered bodies prepared using commercial Si_3N_4 powders.

5.4 SUMMARY

Carbothermal reduction nitriding synthesis has been outlined and the characteristics of powders formed thereby have been discussed. It is often said that raw material powder characteristics have a major influence over sintered body properties in the manufacture of ceramic materials, but numerous aspects of the cause–effect relationship remain unclear. The authors hope to further their understanding of powder characteristics and continue to make improvements, always keeping in mind the question: What is a raw material powder? It would be rewarding if this contribution has been of any use to its readers, and the day when silicon nitride is commonly used as a high-temperature, high-strength material is eagerly awaited.

REFERENCES

1. Komeya, K. & Inoue, H., Synthesis of the α form of silicon nitride from silica. *J. Mat. Sci.*, **10**(7) (1975), 1243–6.
2. Mori, M. & Inoue, H., Synthesis of Si_3N_4 from SiO_2. 1st. Kouon Zairyo Kiso Touronkai, Abstracts. 1981.
3. Mori, M., Komeya, K., Tsuge, A. & Inoue, H., Silicon nitride powder. *Bull. Ceram. Soc. Japan,* **17**(10) (1982), 834–40.
4. Imai, I., Sano, A., Mori, M. & Ishii, T., Synthesis of silicon nitride powder by carbothermal reduction of silica. Nippon Seramikkusu Kyokai Annual Meeting, Abstracts 1986, pp. 239–40.
5. Yamaguchi, A., Influence of oxygen and nitrogen partial pressure on stability of metal, carbide, nitride and oxide in carbon containing refractories. *Refractories*, **38**(4) (1986), 2–11.

6 | Developments in Si₃N₄ Powder Prepared by the Imide Decomposition Method

Y. KOHTOKU

ABSTRACT

Amorphous Si_3N_4 prepared from $Si(NH)_2$ can be converted into three kinds of powders, α-Si_3N_4, β-SiAlON, and β-SiC, as well as into Si_3N_4 whiskers, by varying the crystallization conditions. These powders, which are of equally high purity and are composed of self-crystallized sub-micron particles, are useful as raw materials for engineering ceramics. An Al matrix composite reinforced by Si_3N_4 whiskers exhibits high strength compared with one reinforced by Al alone.

6.1 INTRODUCTION

Ube Industries have brought together numerous original techniques to develop a process of manufacturing silicon nitride powders using imide decomposition: Fig. 6.1 is a flowchart of this process.

Several grades of silicon nitride powder are manufactured, and these are distinguished primarily by specific surface area. The grade can be changed by controlling the conditions of crystallization from amorphous silicon nitride, because this step is referred to as the build-up process.

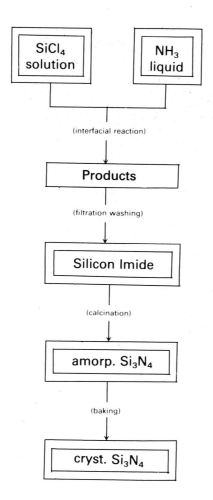

Fig. 6.1. Ube process of Si_3N_4 preparation.

E-10, a typical grade, has extremely few metallic impurities, very high purity, and a high α-phase content; it comprises sub-micron equiaxed particles. This powder is also characterized by an extremely uniform particle size and grain morphology (see Fig. 6.2), and becomes an excellent raw material for silicon nitride ceramics. The characteristics of this powder, and of sintered bodies formed using the powder, are shown in Tables 6.1 and 6.2, respectively. This information was presented in 1983 at the First International Symposium on Ceramic Components in Engines, held in Hakone, Japan.[1]

Fig. 6.2. SEM photomicrograph of Ube SN-E-10.

Chemical analysis	
N	38·7%
O	1·3%
Cl	50 ppm
Fe	50 ppm
Ca	<10 ppm
Al	22 ppm
BET surface area	13 m²/%
Phase composition	
Crystallinity	100%
$\beta/(\alpha+\beta)$	3%

Table 6.1 Typical Properties of Si_3N_4 Powder UBE-SN-E-10

Bulk density	3240 kg/m² (99·0%)
Bending strength, three-point	1000 MPa (R.T.)
Span 30 mm, JIS R-1601-1981	500 MPa (1200°C)
Young's modulus	280 GPa
K_{IC} (CMF method)	7·8 MN/m$^{3/2}$
Vickers microhardness (Load 200 g, 5 s)	17 GPa
Thermal shock fracture resistance	>900°C

Table 6.2 Typical Properties of Sintered Si_3N_4 Made from UBE-SN-E-10

Standard composition: $89·5Si_3N_4 + 5·0Y_2O_3 + 5·5Al_2O_3$ (in wt%).
Condition: 1750°C for 4 h.

Amorphous silicon nitride is an aggregate of extremely small particles (of the order of 100 Å) that exhibits no X-ray diffraction and reacts sensitively to moisture and oxygen, even at room temperature.

The author has attempted to synthesize different materials by mastering this active powder, and has discovered that not only silicon nitride whiskers but also β-SiAlON and β-SiC powders can easily be obtained. This chapter will briefly discuss the preparation method and characteristics of these powders.

In general, β-SiAlON is obtained as a sintered body, using a mixture of silicon nitride, alumina, and aluminum nitride powders, but it has been pointed out that the characteristics of this material have not been fully utilized because of exaggerated grain growth and a poorly distributed composition. In order to resolve these problems, a method has been proposed in which the sintered body is made directly from a β-SiAlON powder.

β-SiC powders are also used as raw materials for sintered bodies, and those with fine grain and an absence of oxygen and free carbon are preferred.

Silicon nitride whiskers are favorable candidates as materials for use in composites such as FRP (fiber-reinforced plastic) and FRM (fiber-reinforced metal).

6.2 EXPERIMENTS

(1) *Amorphous silicon nitride.* Obtained by the thermal decomposition of silicon diimide, this has a specific surface area of approximately 300 m^2/g.

(2) *β-SiAlON powder.* Amorphous silicon nitride, alumina, and aluminum powders were mixed to obtain specified silicon, aluminum, and oxygen contents. β-SiAlON powder was obtained by heat-treating the powder mixture at a temperature of 1200–1700°C in an N_2 atmosphere.

(3) *β-SiC powder.* A carbon crucible containing amorphous silicon nitride was placed in CO and/or CO_2 atmospheres and heat-treated. The transformation into β-SiC occurred at 1500–1700°C.

(4) *Silicon nitride whiskers.* These are obtained by heat-treating amorphous silicon nitride containing trace

amounts of additives such as iron powder. Aluminum matrix FRM was produced by rubber-pressing the silicon nitride whiskers to make a preform, and pressing molten aluminum into the preform using high-pressure casting.

6.3 RESULTS AND DISCUSSION

6.3.1 β-SiAlON Powders

β-SiAlON is represented here by $Si_{6-z}Al_zO_zN_{8-z}$, where z is in the range 0–4·2. Typical results of synthesis using varying z values are shown in Table 6.3. It can be seen that the ratios of the four elements closely coincide with calculated values.

The grain morphology of the powder for $z = 2$ is shown in Fig. 6.3. The X-ray diffraction pattern for this powder showed only a

Table 6.3 Synthesis of β-SiAlON Powder

Intended z value	Raw materials (g)			Observed composition of each element (%) (calc.)			
	Amorphous Si_3N_4	γ-Al_2O_3	Al	Si	Al	O	N
1	137·5	11·8	9·0	49·8 (50·2)	8·6 (9·2)	9·3 (5·5)	35·3 (35·1)
2	127·6	38·8	16·0	38·9 (39·7)	18·7 (19·1)	12·1 (11·3)	30·3 (29·8)
3	82·6	51·1	19·0	28·0 (29·7)	27·3 (28·7)	19·3 (17·0)	25·4 (24·7)

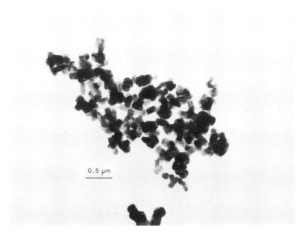

Fig. 6.3. TEM photomicrograph of β-SiAlON powder.

Table 6.4 Relation Between z and Lattice Parameter

z	a (Å) Calculated	a (Å) Observed	c (Å) Calculated	c (Å) Observed
0	7·603	—	2·919	—
1	7·636	7·631	2·937	2·934
2	7·663	7·656	2·963	2·957
3	7·685	7·686	2·995	2·988
4	7·716	—	3·005	—

β-phase. A comparison of the lattice constant obtained from the X-ray diffraction pattern and calculated values,[2] when z was changed, are shown in Table 6.4. It can be seen that the two coincide closely.

These results demonstrate that the synthesized powder is an ideal β-SiAlON. The powder may also be used as a raw material for SiAlON ceramic materials.

6.3.2 β-SiC Powder

SiC synthesis conditions and results are shown in Table 6.5. Although the exact mechanism remains unclear, it has been found that introducing H_2 to the reaction atmosphere is extremely effective in the selective preparation of SiC. It is a characteristic feature of this method that SiC powder containing a minimum amount of oxygen and free carbon can be obtained, without post-treatment. This powder consists of fine tetrahedral particles (see Fig. 6.4).

Table 6.5 Preparation of SiC from Amorphous Si_3N_4

No.	Crystallization conditions						Properties of products				
	Atmosphere (ml/min)				Temperature (°C)	Time (h)	Chemical analysis (%)				Specific surface area (m^2/g)
	CO	H_2	Ar	N_2			Total C	Free C	N	O	
1	100		100		1650	8	28·3	0·4	2·0	0·5	12·1
2	20	20	160		1600	2	28·9	0·4	0·3	0·8	10·5
3	20	20		160	1600	2	29·1	0·5	0·7	0·7	10·8

Si_3N_4 Powder by Imide Decomposition

6.3.3 Silicon Nitride Whiskers

A representative scanning electron micrograph (SEM) and chemical analysis data are provided in Fig. 6.5 and Table 6.6, respectively.

The tips of the whiskers were closely examined to gain information about the crystallization process. Virtually all of the whiskers were broken in post-crystallization handling, but a few had drops on their tips containing large quantities of iron. This

Fig. 6.5. SEM photomicrograph of Ube SN-W.

Table 6.6 Properties of Si₃N₄ Whisker UBE-SN-W

Diameter	0·1–0·4 μm, mainly
Length	5-20 μm
Aspect ratio, L/D	20–100
Density	3·18 g/ml
Crystal type	α type
Fe content	4000–5000 ppm
Ca content	<100 ppm
Al content	<100 ppm
O content	2·0–3·0%

indicates that crystal growth in this method is also a result of the vapor–liquid–solid (VLS) mechanism (see Fig. 6.6).[3] XRD and ED analyses showed that the whiskers comprised an α-phase and grew along the c-axis.

It is believed that this whisker could be used in composites such as FRP, FRM, and FRC. The bending strength of whisker reinforced metals containing 25 vol% dispersed silicon nitride whisker in an industrial pure-aluminum matrix was tested at different temperatures. The results of these measurements and a

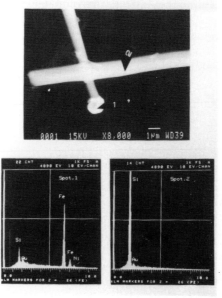

Fig. 6.6. Analysis of Si₃N₄ whisker.

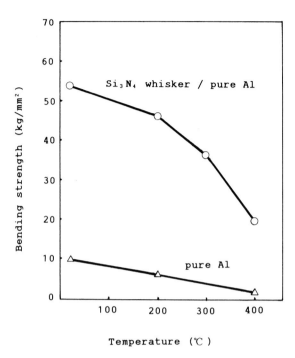

Fig. 6.7. Bending strengths at elevated temperatures of Si_3N_4 whisker (Ube SN-W) reinforced aluminum composites. Matrix pure Al; 25% As cast.

comparison with a matrix of pure aluminum alone are shown in Fig. 6.7.

The reinforcing effect of silicon nitride whiskers is clear. Some problems remain in the uniform dispersion of whiskers in a matrix, but the solution of these should make it possible to obtain even stronger composites.

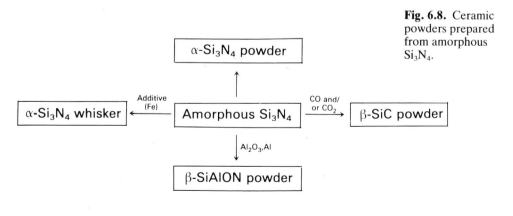

Fig. 6.8. Ceramic powders prepared from amorphous Si_3N_4.

6.4 SUMMARY

It has become clear that it is possible, using amorphous silicon nitride as a raw material, to obtain several types of ceramic raw material powders at relatively low temperatures and with high efficiency, the resulting particles being characterized by high purity and uniform grain morphologies.

The process described in this work is summarized graphically in Fig. 6.8.

REFERENCES

1. Yamada, T., Kawahito, T. & Iwai, T., Preparation of silicon nitride powder by imide decomposition method. In *Proceedings of the 1st International Symposium on Ceramic Components in Engines, 1983, Hakone, Japan*, ed. S. Sōmiya, E. Kanai & K. Ando, KTK Scientific Publishers, Tokyo, 1984 p. 333.
2. Mitomo, M., Kuramoto, N., Tsutsumi, M. & Suzuki, H., The formation of single phase Si-Al-O-N ceramics. *Yogyo-Kyokai-Shi*, **86** (1978) 526.
3. Wagner, R. S. & Ellis, W. C., Vapor-liquid-solid mechanism of single crystal growth. *Appl. Phys. Lett.*, **4** (1984), 89.

7 State of the Art of Silicon Nitride Powders Obtained by Thermal Decomposition of Si(NH)₂ and the Injection Molding Thereof

T. ARAKAWA

ABSTRACT

High-quality Si_3N_4 powder prepared by the thermal decomposition of $Si(NH)_2$ (silicon diimide) is characterized and its injection molding behavior is studied. This powder is characterized by high purity, fine grain, and a narrow particle size distribution. The effect of injection molding on powder characteristics is also investigated. Hot-kneading and hot-fluidity properties of the powder, when mixed with a polymer binder, are dependent on isostatic cold press ($1\cdot 5\, ton/cm^2$) green density. It is thought that injection molding behavior could be modified in order to achieve a higher green density.

7.1 INTRODUCTION

Silicon nitride ceramics are characterized by high-temperature strength and excellent resistance to deformation and corrosion. High fracture toughness is also provided by the crack deflection mechanism and the pull-out effect of the columnar grains. Hence these materials maintain the advantages that are unique to ceramics while at the same time offering some improvement of the brittleness that plagues these materials. Room-temperature strength, high-temperature strength, fracture toughness, hardness, thermal shock resistance, specific gravity and other sintered body characteristics are considerably better balanced than in other ceramic materials. As a result, research and technological development in the area of silicon nitride ceramics are being actively pursued in various fields.

Since silicon nitride is difficult to sinter, sintering is usually performed through the liquid phase with the addition of a sintering aid such as Y_2O_3, Al_2O_3, AlN, or MgO. Sintering properties, and characteristics of the resulting sintered body, may vary considerably depending on the type of sintering aid, making possible the development of a wide variety of silicon nitride ceramics. Like alloys, silicon nitride allows material design based on application and shape.

In order to allow use at high temperatures and to develop superior strength, toughness, and hardness characteristics, further development and improvement of materials (mainly raw material powders and sintering aids), and molding, firing, and processing technologies are necessary. The characteristics of the starting material are particularly important because of the major influence they have over all later processes.

There have been numerous reports on methods of synthesizing silicon nitride powders.[1] This chapter briefly describes synthesis using the thermal decomposition of imides and the characteristics of powders thus fabricated, and offers one example of the improvement of injection molding properties (in particular, kneading and fluidity) as one of the possible molding methods.

7.2 THE IMIDE THERMAL DECOMPOSITION METHOD AND CHARACTERISTICS OF POWDERS PRODUCED THEREBY

7.2.1 Fabrication Method

Current methods of fabricating powders that are being produced or tested industrially as raw materials for silicon nitride ceramics

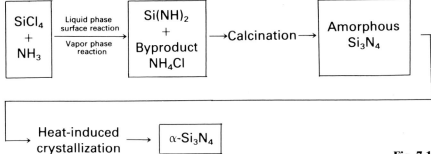

Fig. 7.1. The imide thermal decomposition fabrication process.

include: (1) metallic silicon direct nitriding; (2) silica reduction nitriding; and (3) imide ($Si(NH)_2$) thermal decomposition. High-purity, uniform powders in the submicron range are relatively easy to obtain with the last method.

A typical imide decomposition fabrication process is shown in Fig. 7.1. The process comprises three major steps; (1) imide synthesis, (2) imide thermal decomposition, and (3) crystallization. In the imide synthesis process, silicon tetrachloride ($SiCl_4$) and ammonia (NH_3) are reacted in the vapor[2] or liquid[3] phase to synthesize silicon diimide ($Si(NH)_2$) and ammonium chloride (NH_4Cl). In the following process the $Si(NH)_2$ is thermally decomposed at a temperature of ~1000°C to obtain an amorphous silicon nitride powder. In the final step, this powder is crystallized at a temperature of 1300–1500°C to form α-type silicon nitride powder.

The particle size, α-phase content, and other powder characteristics that are not related to impurities also have some effect on sintering properties and sintered body characteristics, but they play a particularly important role during crystallization, and control of these properties is critical.

7.2.2 Powder Characteristics

Characteristics of currently available commercial 'high-purity' silicon nitride powders are shown in Table 7.1, while a representative transmission electron micrograph is given in Fig. 7.2. The characteristics of these powders can be summarized as

Table 7.1
Characteristics of TOSOH-Si$_3$N$_4$ High-purity Silicon Nitride Powder (grade TS-10)

Typical analysis of chemical purity		*Physical properties of powder*
Fe (ppm)	50	Crystallite size = 400 Å
Al (ppm)	10	Specific surface area = 12 m^2/g
Ca (ppm)	max. 10	α-phase content = 97 wt%
Na (ppm)	max. 10	Green density = 1·55 g/cm^3
K (ppm)	max. 10	(1·5 ton/cm^2 CIP)
O (wt%)	1	
Cl (wt%)	0·03	
C (wt%)	0·1	

follows:

(1) high purity (metallic impurity content of ~100 ppm);
(2) high α-phase content (97%);
(3) small particle size (0·2–0·3 µm) and sharp particle size distribution;
(4) equiaxed grain morphology.

Characteristics generally demanded of silicon nitride powders, and a comparison of imide decomposition powders with those obtained from other methods (in particular, metallic silicon direct nitriding), are shown in Table 7.2.

Concerning characteristic (1), α-phase contents of powders prepared using all methods have been improved to levels exceeding 90%, and there is no significant difference between the methods. Concerning particle size and grain morphology (2, 3), the metallic silicon method employs mechanical grinding, giving its powders an irregular morphology and a wide particle size distribution. In contrast, imide thermal decomposition uses deposition from an amorphous matrix phase, resulting in an equiaxed grain morphology and uniform particle size. Particle size is also small, at 0·2–0·3 µm. Concerning purity (4), since imide decomposition uses a high-purity chemical compound as a raw material there are very few metallic impurities. Molding properties (5) are in general closely related to particle size and particle size distribution (2), and powders produced using the imide decomposition method are difficult to mold. The improve-

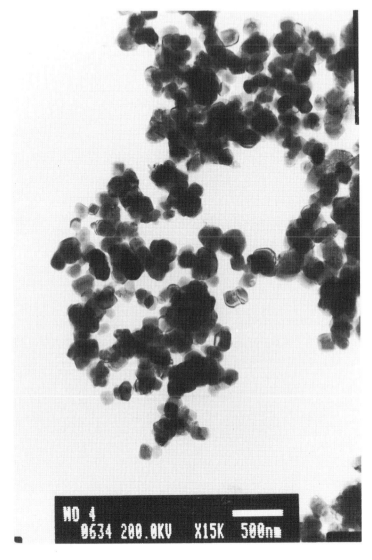

Fig. 7.2. TEM photomicrograph of Si_3N_4 powder, obtained by imide thermal decomposition (grade TS-10).

ment of this will be an important aspect of future work on imide decomposition. The following section describes an attempt at improving molding properties using, as an example, the kneading and fluidity properties of the powder when mixed with a binder during injection molding.

Table 7.2
Characteristics Demanded of Silicon Nitride Powders and the Performance of Current Imide Thermal Decomposition Powders

(1) High α-phase content	—
(2) Small particle size and sharp particle size distribution	○
(3) Equiaxed grain morphology	○
(4) High purity	○
(5) Easy molding	×

7.3 INJECTION MOLDING PROPERTIES OF IMIDE THERMAL DECOMPOSITION POWDERS

It is important to perform injection molding with a small amount of binder because of problems posed by binder removal and contraction. The following types of raw material powders are generally said to be difficult to mold (i.e., they require large amounts of binder):

(1) fine powders (with particle sizes in the sub-micron range);
(2) powders with a plate-like or angular morphology;
(3) powders with a narrow particle size distribution;
(4) powders with high agglomeration energy.

Compared with commercial powders obtained using metallic silicon direct nitriding, powders obtained using imide decomposition are difficult to knead. Turning to the above list in a search for possible factors, we find that the characteristics of both powders for items (1) and (2) are virtually identical, and in some cases imide powders maintain a slight superiority. The main cause for the difference in molding properties would appear to lie in (3) and (4). Sharp particle size distribution (3) is a major characteristic of imide powders and is critical in obtaining sintered bodies with a uniform structure. Consequently, this item was put aside, and agglomeration strength (4) was taken up for study. From the viewpoint of the two fabrication processes (imide thermal decomposition and metallic silicon direct nitriding), the presence or absence of a grinding step appears to have a major influence.

In other ceramics, such as partially stabilized zirconia, a constant linear relation is established between the specific surface

Si(NH)$_2$ Decomposition, Injection Molding

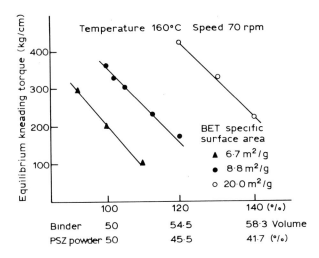

Fig. 7.3. Relation between powder specific surface area, binder content and kneading torque[4] (partially stabilized zirconia).

area (particle size) of the powder and the amount of binder added, as shown in Fig. 7.3.[4] With silicon nitride, however, this relation is not established in most cases. The present author investigated the four powders listed in Table 7.3, which shows the characteristics of each powder. With the exception of the first powder, which was fabricated using metallic silicon direct nitriding, all TOSOH powders have been obtained by imide decomposition, accounting for the virtually identical metallic impurity contents, particle size, and powder morphology. The major difference lies in the green density for isostatic cold-pressing at 1·5 ton/cm^2.

Table 7.3 Characteristics of Si$_3$N$_4$ Powder Used

	α-phase content (wt%)	BET specific surface area (m^2/g)	Green density (g/cm^3)[a]
Direct nitriding powder (D Co.)	90	7	1·83
TS-10	97	13	1·55
A	97	15	1·65
B	97	13	1·73

[a] 1·5 ton/cm^2 CIP.

7.3.1 Method

Hot-kneading properties of the raw material powder and binder (Ceramo IB-27F, Dai-ichi Kogyo Seiyaku Co.) were evaluated at a fixed blend ratio using a Labo Plastomill (Toyo Seiki Co. Ltd) at final torque. Kneading was performed at 50 rpm at a temperature of 150°C. The MFI value of the kneaded object's hot fluidity was also measured with a melt indexer. This was performed according to JIS K-7210 at 150°C with a 10 kg load.

7.3.2 Results

The relation with final torque at each blend ratio is shown in Fig. 7.4, and fluidity is shown in Fig. 7.5. Kneading and fluidity properties correspond closely, and while it is difficult to determine at which boundary regions injection molding is possible, because of changes with molding conditions, final torque during kneading was ~0·5 kg/m, and the MFI value for fluidity was 50 g per 10 min. Evaluations of each powder at the specified blend ratio are shown in Table 7.4. Injection molding of direct nitriding powders was possible at silicon nitride blend

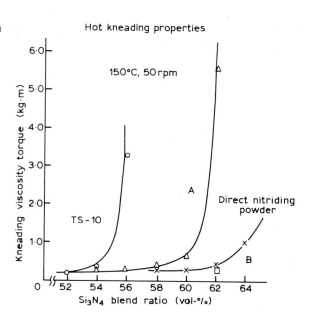

Fig. 7.4. Relation between powder-to-powder blend ratio and kneading torque.

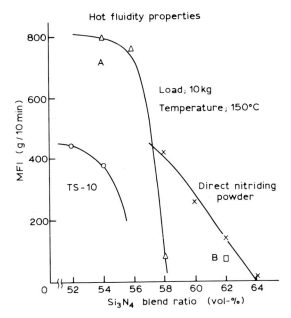

Fig. 7.5. Relation between powder-to-powder blend ratio and fluidity (MFI).

Table 7.4 Evaluations of Powders for Use in Injection Molding

Powder	Blend ratio (vol%)	Kneading properties (Laboplast mill)	Fluidity (melt indexer)	BET specific surface area (m^2/g)	Green density $(g/cm^3)^a$
Direct nitriding powder (D Co.)	58	○	○	7	1·83
	60	○	○		
	62	○	△		
	64	△	×		
TS-10	52	○	○	13	1·55
	54	○	○		
	56	△	×		
A	54	○	○	15	1·65
	56	○	○		
	58	○	△		
	60	△	×		
	62	×	×		
B	62	○	△	13	1·73

^{a} 1·5 ton/cm² CIP.

ratios of up to 62 vol%, while the figure was 54% for TS-10 powder, 58% for powder A, and 62% for powder B. By increasing the CIP green density, the amount of binder required for molding can be reduced, resulting in injection molding properties that are virtually equivalent to those exhibited by direct nitriding powders.

In addition, while the difficulty of kneading imide thermal decomposition powders increased dramatically at a certain point, with a corresponding degradation of fluidity, direct nitriding powders showed a gradual change.

7.3.3 Discussion

Powder characteristics of the three imide powders, with the exception of CIP green density, are virtually identical (i.e., particle size, grain morphology, particle size distribution, impurities, and α-phase content). A linear relation has been established between injection molding kneading properties, fluidity, and CIP green density. As a result, it is believed by the present author that some factor dependent on CIP green density has a major influence on injection molding properties. The fact that CIP green density alone differs while particle size, particle size distribution, and grain morphology remain identical indicates differences in inter-particle friction strength, and the extent of broken inter-particle bridges during pressing, and is believed to indicate the agglomeration energy of the powders. Inter-particle agglomeration energy, therefore, controls the injection molding properties of powders fabricated using imide thermal decomposition, and by reducing this strength injection molding properties equivalent to those exhibited by direct nitriding powders can be obtained. Finally, the fact that the relation of kneading properties and fluidity to binder volume changes drastically at a certain blend ratio for imide powders, while for direct nitriding powders the change is gradual, is believed to reflect differences in the particle size distributions of the two powder types.

7.4 SUMMARY

The characteristics of powders fabricated using imide thermal decomposition, compared to those of powders produced using other methods, are as follows:

Advantages: high purity
uniform particle size, particle size distribution, and grain morphology
Disadvantage: difficult molding

While sintered bodies obtained from such powders are dense and have a uniform microstructure, handling during molding is difficult. Although this is naturally true for injection molding as well, it is possible to produce powders with injection molding properties that are the same as those of direct nitriding powders by reducing inter-particle agglomeration energy.

Silicon nitride development work is currently making the transition from research and development using testpieces to prototypes and eventual commercialization, and it is believed by the author that a detailed study by powder molding methods will become increasingly important.

REFERENCES

1. Arakawa, T., Ohno, K. & Ueda, K., Present status of silicon nitride powder. *Ceramics*, **22**(1) (1987), 34–9.
2. Japanese Patent No. Kokai Tokkyo Koho JP 54-124898 (79-124898), TOSOH Co., Japan (1978).
3. Japanese Patent No. Kokai Tokkyo Koho JP 54-145400, UBE Industries Ltd., Japan, (1978).
4. Saitou, K., *Molding and Polymer Material of Fine Ceramics*. CMC Co., (1985), p. 366.

8 Synthesis of Ultrafine Si_3N_4 Powder Using the Plasma Process and Powder Characterization

N. KUBO, S. FUTAKI & K. SHIRAISHI

ABSTRACT

An ultrafine Si_3N_4 powder was synthesized by the vapor phase reaction of $SiCl_4$ and NH_3 in a thermal plasma arc jet. The as-prepared powder was white and amorphous, and had a mean particle size of 30–40 nm. The production rate was 200 g/h. Heat treatment above 1380°C reduced the specific surface area and increased crystallinity, implying that grain growth and crystallization occurred simultaneously. The oxygen content was reduced to 1 wt% at 1450°C, and the powder thus obtained had a crystallized hexagonal grain with a mean size of 0·2 µm. X-ray diffraction analysis indicated the presence of more than 95% α-phase Si_3N_4, with the remainder being β-phase Si_3N_4. Sintering at 1700°C with the addition of 5 wt% Al_2O_3 and 5 wt% Y_2O_3 as sintering aids resulted in a relative density of 98%, a Vickers hardness of 16 GN/m^2, and a fracture toughness of 7 $MN/m^{3/2}$.

8.1 INTRODUCTION

Silicon nitride (Si_3N_4) sintered bodies are leading candidates for engineering ceramics because of their high strength and resistance to heat, corrosion, and wear, and in some cases have reached the stage of practical application. Since the fabrication of highly functional, highly reliable products requires appropriate control over the microstructure of the sintered body, the Si_3N_4 raw material powder should be of high purity and fine grain.[1-4]

The numerous methods of synthesizing Si_3N_4 powders include: metallic silicon nitriding;[5,6] carbothermal reduction, in which silica is reduced and nitrided with carbon in flowing nitrogen;[7] vapor-phase synthesis, which uses silicon tetrachloride and ammonia at high temperatures;[8,9] and thermal decomposition of silicon diimide.[10-14] In general, since the vapor-phase method allows the use of high-purity gases as a starting material, high-purity, fine-grained powders can be obtained.

Synthesis of ultrafine Si_3N_4 powders by the thermal plasma CVD method can be performed using a d.c. plasma,[15] a radio-frequency plasma,[16-18] or a hybrid plasma.[19-20]

The present authors succeeded in obtaining an amorphous ultrafine Si_3N_4 powder at a stable production rate of ~200 g/h, using the vapor-phase method with a d.c. arc plasma, which allows mass production, and using $SiCl_4$ and NH_3 as starting materials.

The relation of the crystallization of Si_3N_4 powder and the accompanying grain growth to heat treatment was clarified, as was the behavior of Si_3N_4 particles during crystallization, and the synthesis of an Si_3N_4 powder comprising fine hexagonal plate-like particles was made possible. In addition, Al_2O_3 and Y_2O_3 were added to these powders as sintering aids, and their sintering behavior was studied.

8.2 EXPERIMENTS

8.2.1 Experimental Procedure

A schematic diagram of the reactor used for synthesis is shown in Fig. 8.1. The reaction system was a 1-atm flowing system substituted with argon, and argon gas was used as the arc jet gas to ignite the d.c. arc plasma: the output of the latter was ~6 kW. $SiCl_4$ (99·9% pure) was introduced into the d.c. plasma torch by a mass-flow controller, with argon as a carrier gas. The $SiCl_4$ was

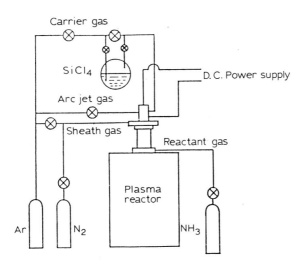

Fig. 8.1. Schematic diagram of plasma CVD apparatus.

introduced into the plasma flame where it underwent thermal decomposition, chemically reacted with NH_3, and experienced rapid cooling to form an ultrafine Si_3N_4 powder. NH_3 was supplied in excess (with respect to $SiCl_4$) such that the $NH_3:SiCl_4$ molar ratio was varied from 1·3 to 2·7, and the resulting yield was sought. Reaction conditions are shown in Table 8.1.

In order to remove by-products such as NH_4Cl and HCl from the ultrafine powder thus obtained, calcining was performed for 4–10 h at a temperature of 500–1000°C in flowing nitrogen. Due to the amorphous nature of the ultrafine powder and its resulting susceptibility to oxidation, heat treatment was required. By firing the powder for 2 h in the temperature range 1200–1550°C in flowing nitrogen, crystallization and grain growth were induced.

Furthermore, 5 wt% Al_2O_3 (99·99% pure) and 5 wt% Y_2O_3

Table 8.1 Experimental Conditions for Preparation of Ultrafine Si_3N_4 Powder

DC power supply	6 kW
Plasma gas, Ar	10 l/min
$SiCl_4$ feed rate	13 g/min
Carrier gas, Ar	4 l/min
Reactant gas, NH_3	2·2–4·6 l/min
$NH_3:SiCl_4$ molar ratio	1·3–2·7

(99·9% pure) were added as sintering aids to both the amorphous and the crystallized Si_3N_4 powders, which were then ball-milled. After pressing into a pellet in a die, rubber-press molding at 2 ton/cm^2 was performed. The pellet was placed in a BN-lined graphite crucible and embedded in the sintering bed. Sintering was performed in a Tammann furnace for 1 h at a temperature of 1550–1700°C in flowing nitrogen with a heating rate of 10°C per minute. The sintering bed consisted of a mixture of Si_3N_4 (40 wt%), Al_2O_3 (5 wt%), Y_2O_3 (5 wt%), and BN (50 wt%).

8.2.2 Evaluation

Various tests were conducted on the ultrafine Si_3N_4 powder obtained using the plasma CVD method and the Si_3N_4 powders that had undergone crystallization and grain growth: identification of the formation phase using an X-ray powder diffractometer (Cu–K_α); the measurement of crystallization;[21] crystallite size and BET specific surface area; bonding conditions (by infra-red absorption using KBr tablets); determination of the composition using chemical analysis; and observation by transmission electron microscopy.

Chemical analysis for the various elements was carried out using weight analysis (Si), the quantitative method of NaOH titration (N), inert gas extraction (O), vacuum fusion (H), and ion chromatography (Cl).

The sintered bodies thus obtained were polished with #600, #800, and #1200 wheels and diamond paste after parallel grinding. The Archimedes liquid displacement technique was used to calculate the density of the sintered body, and the Vickers hardness and fracture toughness, K_{IC},[22] were also measured.

8.3 RESULTS AND DISCUSSION

The relation between the yield of the Si_3N_4 powder obtained using a d.c. plasma and the $NH_3:SiCl_4$ molar ratio is shown in Fig. 8.2. Regardless of the molar ratio, only amorphous Si_3N_4,

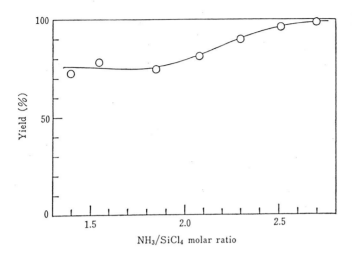

Fig. 8.2. Yield vs. $NH_3:SiCl_4$ molar ratio.

NH_4Cl, and HCl were formed. At molar ratios of 1·4 to 2·0 the yield was approximately 80%, while at molar ratios of more than 2·3 the yield rose above 90%, but was accompanied by an increase in the NH_4Cl by-product. When the molar ratio dropped below 1·3 there was a drastic reduction in yield. Therefore, from the standpoints of reaction yield and by-product processing, a molar ratio range of 1·5 to 2·0 was chosen for synthesis.

The powder thus obtained was a white and amorphous powder (see Fig. 8.3) showing a halo pattern at $2\theta(Cu-K_\alpha) = 20-40°$ according to X-ray diffraction analysis. It also showed broad infra-red absorption near $1000\,cm^{-1}$ due to Si–N bonding. No SiO absorption was observed (Fig. 8.4). Absorption due to N–H bonding was shown at $3400\,cm^{-1}$.

Figure 8.5 is a transmission electron micrograph showing that the powder consists of semi-isotropic ultrafine particles with a particle size of 30–40 nm. The BET specific surface area was $\sim 60\,m^2/g$. Since this ultrafine powder is amorphous, it is susceptible to oxidation and decomposition in air, resulting in an increased oxygen content,[12,20] and also the green density of the formed body tends to be low. As a result, it was decided that material improvements were required in order to make the powder suitable for use as a sintering raw material, and crystallization and grain growth were induced by heat treatment.

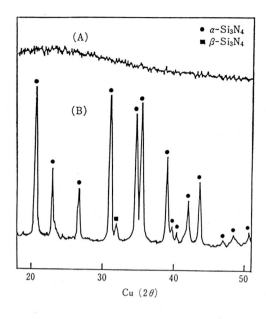

Fig. 8.3. X-ray diffraction patterns of Si_3N_4. (A) Si_3N_4 obtained by the plasma CVD method; (B) commercial Si_3N_4.

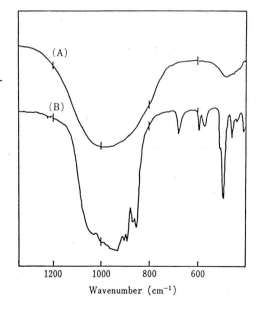

Fig. 8.4. IR absorption spectra of Si_3N_4. (A) Si_3N_4 obtained by the plasma CVD method; (B) heat-treated Si_3N_4 at 1430°C for 2 h in N_2.

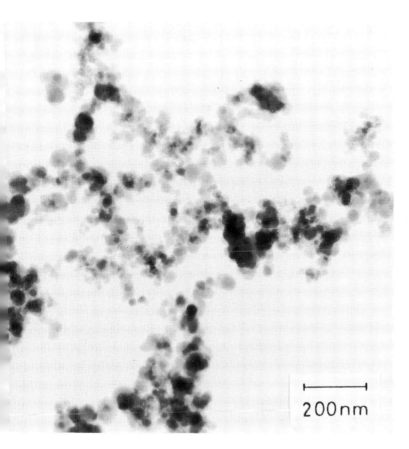

Fig. 8.5. TEM photomicrograph of ultrafine Si_3N_4 powder made by the plasma process.

The relation between calcining temperature and specific surface area when amorphous Si_3N_4 powder obtained by plasma CVD is heated is shown in Fig. 8.6. From room temperature to 1200°C, specific surface area was unchanged. From 1300°C it began to decrease, and a rapid drop occurred at 1380–1400°C. The specific surface area was virtually constant at temperatures above 1450°C. A reduction in the specific surface area indicates the growth of Si_3N_4 particles.

The relation between firing temperature and the crystallization of the Si_3N_4 powder is shown in Fig. 8.7. Up to 1200°C, there was only a halo pattern at $2\theta(Cu-K_\alpha) = 20-40°$, and the powder remained in an amorphous state. From 1300°C broad and weak diffraction lines were observed in the Si_3N_4, and crystallization

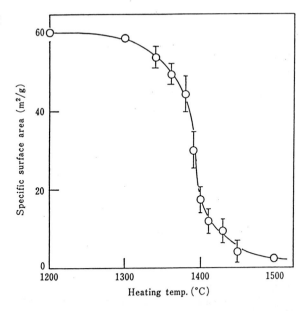

Fig. 8.6. Effect of heating temperature on specific surface area.

increased together with temperature. However, α-Si_3N_4 and β-Si_3N_4 ratios remained constant (α-type >95%) regardless of the firing temperature.

Using X-ray diffraction analysis, the crystallite size was found to be approximately 30 nm at 1300°C. The size then increased with firing temperature up to ~1400°C, where it increased

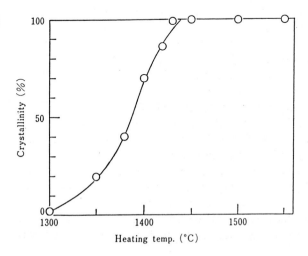

Fig. 8.7. Effect of heating temperature on crystallinity of Si_3N_4.

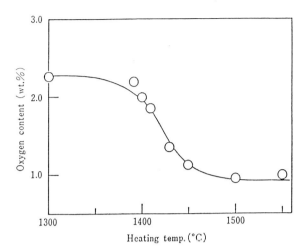

Fig. 8.8. Effect of heating temperature on oxygen content.

dramatically, growing to ~70 nm at a temperature of 1450°C. These results suggest that crystallization does not proceed as the result of nucleus generation and crystallite growth within the various amorphous particles but rather occurs simultaneously with grain growth.

The effect of firing temperature on the Si_3N_4 powder oxygen content is shown in Fig. 8.8. Si_3N_4 powder obtained with plasma CVD is susceptible to moisture adsorption and oxidation, and normally contains 3–5 wt% oxygen. When this amorphous powder was fired, a drop in oxygen content was observed from ~1400°C. At temperatures above 1450°C oxygen content had dropped to ~1 wt%. This phenomenon coincides with the behavior of amorphous Si_3N_4 reported by Abe et al.[23] The reduction in oxygen content is believed to be due to the fact that the oxide film present on the surface of Si_3N_4 particles reacts chemically and vaporizes during the grain growth process.

The characteristics of Si_3N_4 powder obtained by firing at 1430°C are shown in Table 8.2. Figure 8.9 is a transmission electron micrograph of crystallized Si_3N_4 powder. The crystallized particles were observed to be uniform and had a particle size of ~0·2 µm. The crystallites had a hexagonal grain morphology, and no acicular crystals were observed.

The relation between green density and crystallinity of two types of rubber-pressed bodies, one with a powder mixed with

Table 8.2 Characteristics of Si$_3$N$_4$ Powder Calcined at 1430°C for 2 h in N$_2$

Grain size	0·2 µm
Specific surface area	10·6 m^2/g
Chemical analysis	
N	38 wt%
O	1·3 wt%
Cl	100 ppm
Crystallinity	100%
α-Si$_3$N$_4$ phase	>95%
Grain morphology	Equiaxed

Fig. 8.9. TEM photomicrograph of Si$_3$N$_4$ powder calcined at 1430°C.

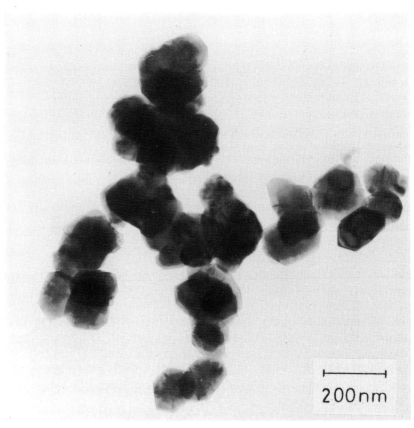

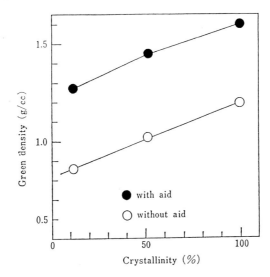

Fig. 8.10. Dependence of green density on crystallinity.

sintering aids in a ball mill, and the other with a powder free from aids and without ball milling, is shown in Fig. 8.10. Both of the powders exhibited a linear relation between crystallinity and green density. It can be seen that, in order to increase the green density any further, 100% crystallization is favorable.

Sintering was performed with the addition of 5 wt% Al_2O_3 and 5 wt% Y_2O_3 as sintering aids. When amorphous and low-crystallinity powders were used, there was a high rate of contraction, and in some cases cracks appeared. The relation between the bulk density of sintered bodies using three 100%-crystallized powders and sintering temperature is shown in Fig. 8.11. Calcining temperatures for the three powders were 1430°C, 1450°C, and 1500°C. Mean particle sizes were 0·2, 0·3, and 0·5 μm, respectively. The powder calcined at 1430°C showed a relative density of 98% at sintering above 1650°C. The powder calcined at 1500°C, on the other hand, exhibited inferior sinterability. The powder calcined at 1450°C exhibited behavior falling between these two. Despite these differences, all three powders showed virtually equal bulk densities (3·19 g/cm³) at sintering temperatures of 1700°C. The differences in sintering behavior are believed to stem from differences in Si_3N_4 powder particle size. The room-temperature mechanical characteristics of

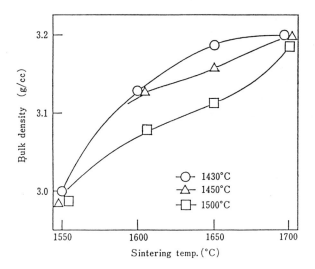

Fig. 8.11. Dependence of bulk density on sintering temperature.

a sintered body using a powder calcined at 1430°C and sintered at 1700°C were a Vickers hardness of 16 GN/m² and a fracture toughness (K_{IC} value) of 7 MN/m$^{3/2}$.

8.4 SUMMARY

An amorphous Si_3N_4 powder was stably synthesized at a production rate of ~200 g/h using the plasma CVD method with $SiCl_4$ and NH_3 as starting materials. The product thus obtained was white, and was determined from X-ray diffraction and infra-red absorption to be amorphous. The powder consisted of ultrafine particles with a specific surface area of 60 m²/g and a mean particle size of 30–40 nm. The characteristics of this amorphous Si_3N_4 powder were as follows:

(1) Grain growth and crystallization proceeded simultaneously when heat treatment was performed, and at temperatures above 1380°C significant reductions in specific surface area and increases in crystallinity could be observed.
(2) The oxygen content began to drop at ~1400°C, reaching a constant value of ~1 wt% at temperatures above 1450°C.
(3) Using 1430°C heat treatment, fine-grained hexagonal

plate-like particles with 100% crystallization, an α-content higher than 95%, and a mean particle size of 0·2 μm were obtained.

(4) A sintered body with a relative density of 98% was obtained with 1700°C sintering, using a sintering aid. The Vickers hardness and fracture toughness value, K_{IC} (at room temperature), were $16\,GN/m^2$ and $7\,MN/m^{3/2}$, respectively.

REFERENCES

1. Richerson, D. W., Effect of impurities on the high temperature properties of hot-pressed silicon nitride. *Am. Ceram. Soc. Bull.*, **52** (1973) 560–69.
2. Giachello, A., Martinengo, P. & Tommasini, G., Sintering and properties of silicon nitride containing Y_2O_3 and MgO. *Am. Ceram. Soc. Bull.*, **59** (1980), 1212–15.
3. Tsuge, A., Nishida, K. & Komatsu, M., Effect of crystallizing the grain-boundary glass phase on the high-temperature strength of hot-pressed Si_3N_4 containing Y_2O_3. *J. Am. Ceram. Soc.*, **58** (1975) 323–6.
4. Gazza, G. E., Hot-pressed Si_3N_4. *J. Am. Ceram. Soc.*, **56** (1973), 662.
5. Messier, D. R., Wong, P. & Ingram, A. E., Effect of oxygen impurities on the nitridation of high-purity silicon. *J. Am. Ceram. Soc.*, **56** (1973), 171–2.
6. Campus-Loriz, D., Howlett, S. P., Riley, F. L. & Yusaf, F., Fluoride accelerated nitridation of silicon. *J. Mat. Sci.*, **14** (1979), 2325–34.
7. Komeya, K. & Inoue, H., Synthesis of the α form of silicon nitride from silica. *J. Mat. Sci.*, **10** (1975), 1243–6.
8. Kato, A., Ono, Y., Kawazoe, S. & Mochida, I., Finely divided silicon nitride by vapor phase reaction between silicon tetrachloride and ammonia. *Yogyo–Kyokaishi (J. Ceram. Soc. Japan)*, **80** (1972), 114.
9. Aboaf, J. A., Some properties of vapor deposited silicon nitride films obtained by the reaction of $SiBr_4$ and NH_3. *J. Electrochem. Soc.*, **116** (1969), 1736–40.
10. Glemser, O. & Nauman, P., Über den thermischen abbau von siliciumdiimid $Si(NH)_2$. *Z. Anorg. Allgem. Chem.*, **298** (1959), 131–41.
11. Billy, M., Preparation and definition of silicon nitride. *Ann. Chim.*, **4** (1959), 795–851.
12. Madiyasni, K. S. & Cooke, C. M., Synthesis, characterization, and

consolidation of Si_3N_4 obtained from ammonolysis of $SiCl_4$. *J. Am. Ceram. Soc.*, **56** (1973), 628–33.
13. Morgan, P. E. D., Final Annual Report, A-C3316, Contract N00014-72-C-0262, Franklin Institute Research Laboratories, Philadelphia, 1973.
14. Yamada, T., Kawahito, T. & Iwai, T., *J. Mat. Sci. Lett.*, **2** (1983), 275–8.
15. Kubo, N., Futaki, S., Shiraishi, K. & Shimizu, M., Synthesis of ultrafine Si_3N_4 powder by plasma process and powder characterisation. *Yogyo-Kyokaishi (J. Ceram. Soc. Japan)*, **95** (1987) 59.
16. Chin-Wen Zhu & Jia-Peng Yan, Preparation of ultrafine Si_3N_4 powders in radio-frequency plasma. In *Symp. Proc. 7th Int. Symp. Plasma Chem.*, (1985), pp. 657–61.
17. Vogt, G. L., Vigil, R. S., Newkirk, L. R. & Trkula, M., Synthesis of ultrafine ceramic and metallic powders in a thermal argon RF plasma. In *Symp. Proc. 7th Int. Symp. Plasma Chem.*, (1985), pp. 668–73.
18. Koh, E., Yamasaki, Y., Ka, H., Saito, T. & Yoda, K., Ultrafine α-Si_3N_4 powder prepared by RF plasma. In *Yogyo–Kyokai Nenkai (Proceedings of Annual Meeting, 1983)*. Ceramics Society Japan, (1983), p. 123.
19. Yoshida, T., Tani, T., Nishimura, H. & Akashi, K., Characterization of a hybrid plasma and its application to a chemical synthesis. *J. Appl. Phys.*, **51** (1983) 640–6.
20. Futaki, S., Shiraishi, K., Shimizu, T. & Yoshida, T., Synthesis and characterization of ultrafine silicon nitride powder produced by a hybrid plasma technique. *Yogyo–Kyokaishi (J. Ceram. Soc. Japan)*, **94** (1986), 7.
21. Yamada, T., Masunaga, K., Kunisawa, T. & Kotoku, Y., Measuring method of crystallinity of silicon nitride powder. *Yogyo-Kyokaishi (J. Ceram. Soc. Japan)*, **93** (1985), 64.
22. Niihara, K., Indentation microfracture of ceramics—its application and problems. *Ceramics Japan (Bull. Ceram. Soc. Japan)*, **20**, (1985), 12.
23. Abe, O., Kanzaki, S. & Tabata, H., De-oxygen treatment of silicon nitride raw powder. *24th Yogyo Kisotoronkai (Proceedings of the 24th Symposium of the Basic Science of Ceramics)*, Vol. 1, 1986.

9 | The Influence of Si_3N_4 Powder Characteristics on Sintering Behavior

K. ICHIKAWA

ABSTRACT

The sintering behavior of several kinds of Si_3N_4 powders produced by the silicon nitridation method was investigated in relation to their powder characteristics. The results indicated that the sinterability of the powders was apt to be improved with the metallic impurities content, the specific surface area, and the green density of the compact.

9.1 INTRODUCTION

There have recently been numerous attempts to use ceramics as high-temperature structural materials, and particularly high expectations have been placed on silicon nitride. Since most ceramic manufacturing processes begin with the forming of a raw material powder, production of a good sintered body presupposes the use of a good raw material powder. When the development of silicon nitride ceramics had just begun, only a few varieties of silicon nitride powder were available from which to choose. Recently, however, the number of available raw material powders has increased, high-performance ceramics are increasingly in demand, and the improvement of raw material powder selection has come to be recognized as necessary. Powders are evaluated mainly on the basis of sintering properties, with the question being what kind of characteristics powders should have for use as sintering raw materials. According to research conducted thus far,[1,2] the common powder characteristic factors having an influence on silicon nitride sintering properties are the specific surface area and the impurity content: powders with high metallic impurity or oxygen contents and those with a large specific surface area are easily densified. It has been reported that powders with good packing characteristics,[1] powders with small crystallites, and amorphous powders with a high oxygen content obtained using the vapor-phase method have good sintering properties, while high carbon-content powders synthesized using carbothermal reduction have inferior sintering properties.

In this chapter, sintering tests performed using two silicon nitride powders, one fabricated by Showa Denko using silicon direct nitriding and one produced by another firm, are reported. The influence of powder characteristics on sintering behavior was also studied.

The powder obtained using silicon direct nitriding was fabricated by heating and reacting with microground silicon (with a particle size of less than several tens of micrometers) at a temperature of ~1400°C in a nitrogen atmosphere, and then finely grinding the massive silicon nitride thus formed.

9.2 EXPERIMENTAL METHOD

9.2.1 Methods of Measuring Powder Characteristics

The following techniques were used to measure the α-content (α-phase Si_3N_4); chemical composition, particle size, specific surface area, and apparent density of the silicon nitride powders.

Powder X-ray diffraction patterns were used to measure the diffraction line height of α-Si_3N_4 at $2\theta(Cu-K_\alpha) = 31·0°$ (I_α) and β-Si_3N_4 at 33·7° (I_β). Given these two values, the α-content was calculated from the following equation:[3]

$$\alpha\text{-content}\,(\%) = (-0{\cdot}4434 R_\alpha^2 + 1{\cdot}4434 R_\alpha) \times 100$$

where $R_\alpha = I_\alpha/(I_\alpha + I_\beta)$.

Chemical composition was deduced as follows. Metal impurities were analyzed using an inductive high-frequency plasma (ICP); and oxygen content by using impulse furnace extraction; while measurements of carbon content were based upon JIS-G 1211.

The specific surface area (BET) was measured using nitrogen gas adsorption.

The particle size was measured using laser diffusion.

The loose apparent density was calculated by dropping the powder sample from a height of ~239 mm by vibration, passing a 710-μm mesh sieve through a 100 ml container, and scraping off the excess powder with a blade. The weight of the powder in the container was then measured. For the tapped apparent density, a 100-ml container to which a cap had been attached was filled with the sample powder and tapped 180 times during a 180-s period; the cap was then removed, and the excess powder on the container was scraped off with a blade and measured. The green compact apparent density was calculated by placing 15·0 g of powder sample in a 30-mm diameter die, applying a pressure of 1 ton/cm² for 30 s, and then measuring the height of the green compact with a dial indicator.

9.2.2 Sintering Tests

Sintering aids consisting of 1–3 wt% Al_2O_3 and 1–6 wt% Y_2O_3 were added to the silicon nitride powders, and the powders were ball-mixed and dried to form a granulated powder. This was then placed in a 50-mm diameter die, pressed to a thickness of ~5 mm at a pressure of 300 kg/cm², and pressure-formed by 2 ton/cm² rubber-pressing. The compacts were placed in BN containers and sintered for 4 h at 1650–1800°C in a 10 kg/cm² N_2 atmosphere. The density of the resulting sintered bodies was measured using the Archimedes liquid displacement technique. In addition, some of the sintered bodies were segmented into 3 × 4 × 35 mm pieces and polished, and the bending strength was measured according to JIS-R1601.

9.3 RESULTS AND DISCUSSION

9.3.1 Powder Characteristics

The powder characteristics of each sample material are shown in Table 9.1. Looking at the impurity content, it can be seen that NU-10, NU-11, and NU-12 (collectively referred to as the NU-10 series) contained relatively large amounts of Fe and Al, in contrast to NU-30 and A, which contained few metallic impurities. The mean particle size for powders in the NU-10 and NU-20 series exceeded 1 µm, while NU-30 and A had mean sizes of just under 1 µm. The relation between particle size and specific surface area was, at least for the powders examined here, virtually constant. Ordinarily, powders with a high apparent density offer easy sintering and a low sintering contraction rate, making them well suited for use as sintering raw materials. Powder packing properties can be easily observed during molding, and in general powders with large particle sizes, wide particle size distributions, and spherical grain morphologies offer good packing properties. For powders in the NU-10 and NU-20 series, the apparent density increases with the particle size. NU-30, fabricated on the basis of a study of fine grinding methods and conditions, showed unique characteristics and exhibited better packing properties than any of the other powders. However, it remains uncertain just how much significance powder packing properties have during sintering. When silicon nitride powders are used as sintering raw materials, a sintering aid is usually wet-mixed, and it is believed that powder characteristics may change during this process. It was therefore decided to investigate the relation between the green density of powders prepared with the addition of a sintering aid and the apparent density of undoped powders. As shown in Fig. 9.1, there was a linear relation between the two for virtually all powders, so it can be safely assumed that, at least during ordinary ball-mixing, powder packing properties do not change. Powder A, despite a low green compact apparent density, has a high green density. This is believed to be due to the high agglomeration energy of the powder particles.

Table 9.1
An Overview of Silicon Nitride Powder Characteristics

Sample No.	α-content (%)	Chemical composition (wt %)					Specific surface area (m^2/g)	Mean particle size (μm)	Apparent density		
		Fe	Al	Ca	O	C			Loose	Tapped	Green compact
NU-10	91	0·36	0·15	0·04	1·9	0·07	10·0	1·25	0·36	0·72	1·72
NU-11	87	0·36	0·17	0·04	1·8	0·09	8·5	1·42	0·41	0·75	1·75
NU-12	77	0·37	0·20	0·04	1·7	0·08	6·8	1·67	0·44	0·81	1·82
NU-20	93	0·08	0·09	0·004	1·8	0·07	9·8	1·23	0·34	0·71	1·70
NU-21	87	0·10	0·10	0·003	1·8	0·08	8·7	1·45	0·38	0·74	1·73
NU-30	95	0·04	0·03	0·004	1·6	0·25	12·4	0·97	0·60	1·12	1·90
A	96	0·06	0·06	0·03	1·6	0·17	11·5	0·95	0·28	0·61	1·75

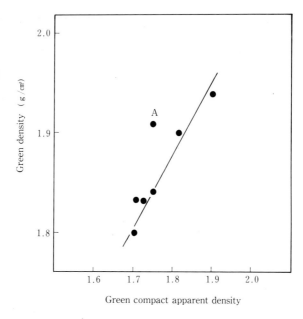

Fig. 9.1. Green density versus green compact apparent density for various silicon nitride powders.

9.3.2 Sintering Properties

The relation between the sintering temperature and the bulk density, for the case when 6 wt% Y_2O_3 and 2 wt% Al_2O_3 were added as sintering aids, is shown in Fig. 9.2. Given the proper sintering temperature, any powder will densify to a value approaching the theoretical density, but as the sintering temperature drops major differences in sinterability develop between the powders. One reason for the ample densification of the NU-10 series powders at the relatively low temperature of 1650°C is their high metallic impurity content. The stable high densification of NU-30, which has particularly few impurities, in the wide temperature range 1700–1800°C is believed to be due to its large specific surface area and high packing density. Concerning the relation between the raw material α-content and sintering properties, a comparison of the powders in the NU-10 and NU-20 series (with identical impurity contents but differing α-contents) revealed absolutely no difference in bulk densities. Given α-contents similar to those of the powders used here, therefore, the α-content is not believed to influence sintering

Table 9.1
An Overview of Silicon Nitride Powder Characteristics

Sample No.	α-content (%)	Chemical composition (wt%)					Specific surface area (m²/g)	Mean particle size (μm)	Apparent density		
		Fe	Al	Ca	O	C			Loose	Tapped	Green compact
NU-10	91	0·36	0·15	0·04	1·9	0·07	10·0	1·25	0·36	0·72	1·72
NU-11	87	0·36	0·17	0·04	1·8	0·09	8·5	1·42	0·41	0·75	1·75
NU-12	77	0·37	0·20	0·04	1·7	0·08	6·8	1·67	0·44	0·81	1·82
NU-20	93	0·08	0·09	0·004	1·8	0·07	9·8	1·23	0·34	0·71	1·70
NU-21	87	0·10	0·10	0·003	1·8	0·08	8·7	1·45	0·38	0·74	1·73
NU-30	95	0·04	0·03	0·004	1·6	0·25	12·4	0·97	0·60	1·12	1·90
A	96	0·06	0·06	0·03	1·6	0·17	11·5	0·95	0·28	0·61	1·75

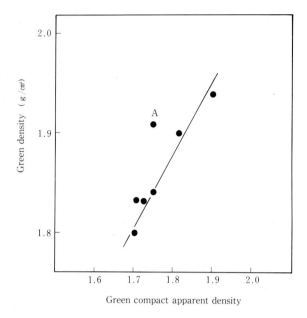

Fig. 9.1. Green density versus green compact apparent density for various silicon nitride powders.

9.3.2 Sintering Properties

The relation between the sintering temperature and the bulk density, for the case when 6 wt% Y_2O_3 and 2 wt% Al_2O_3 were added as sintering aids, is shown in Fig. 9.2. Given the proper sintering temperature, any powder will densify to a value approaching the theoretical density, but as the sintering temperature drops major differences in sinterability develop between the powders. One reason for the ample densification of the NU-10 series powders at the relatively low temperature of 1650°C is their high metallic impurity content. The stable high densification of NU-30, which has particularly few impurities, in the wide temperature range 1700–1800°C is believed to be due to its large specific surface area and high packing density. Concerning the relation between the raw material α-content and sintering properties, a comparison of the powders in the NU-10 and NU-20 series (with identical impurity contents but differing α-contents) revealed absolutely no difference in bulk densities. Given α-contents similar to those of the powders used here, therefore, the α-content is not believed to influence sintering

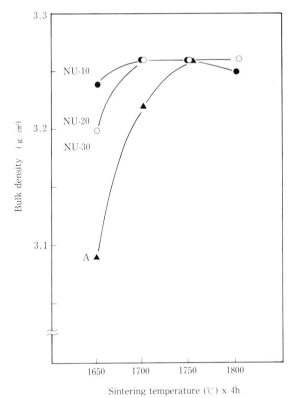

Fig. 9.2. Sintering temperature versus bulk density when 2 wt% Al_2O_3 and 6 wt% Y_2O_3 are added (10 kg/cm^2 in N_2).[4]

properties. However, the same cannot be said of sintered body characteristics. When the bending strength of these testpieces was measured, sintered bodies obtained using powders with a high α-content showed higher strength (see Fig. 9.3).

Next, with the objective of studying sintering properties from a different perspective, an experiment was conducted to determine how the bulk density of certain powders changed when the amount of sintering aid added was varied. The amount of Al_2O_3 added was fixed at 2 wt%, and the amount of Y_2O_3 added was varied from 1 to 6 wt%: the results are shown in Fig. 9.4. It is interesting to note that as the Y_2O_3 volume was reduced the bulk density of A showed a major decrease, while the bulk density of NU-30 showed little change. The reason for this difference in

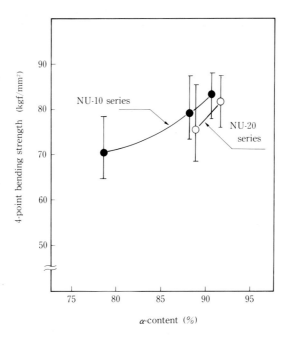

Fig. 9.3. Relation between silicon nitride powder α-content and room-temperature bending strength of the sintered body (sintering conditions— 2 wt% Al_2O_3 + 6 wt% Y_2O_3 sintering aid, 10 kg/cm² in N_2 at 1750°C for 4 h).

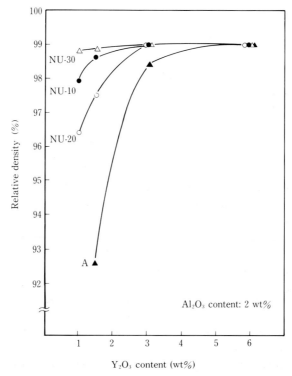

Fig. 9.4. Change in bulk density when the Y_2O_3 content is varied (sintering conditions— 10 kg/cm² in N_2 at 1750°C for 4 h).[4]

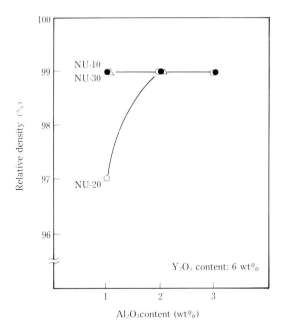

Fig. 9.5. Change in green density when the Al_2O_3 content is varied (sintering conditions— 10 kg/cm² in N_2 at 1750°C for 4 h).

sintering properties, despite virtually identical chemical compositions and particle size, is perhaps due to differences in packing properties. Grain morphology, particle size distribution, and agglomeration energy appear to be more important than mean particle size, and it is believed that these differences influenced sintering properties, particularly when the amount of sintering aid added is small. When the amount of Y_2O_3 added was fixed at 6 wt% and the amount of Al_2O_3 varied from 1 to 3 wt%, there was absolutely no change in the bulk densities of NU-10 and NU-30 (see Fig. 9.5).

9.4 SUMMARY

In this chapter the relation between powder characteristics and sinterability for silicon nitride powders prepared using silicon direct nitriding was investigated, and the following points became clear:

(1) Powders with large specific surface areas offer easy sintering.
(2) Metallic impurities promote sintering. At low temperatures in particular they have a major effect on sintering.
(3) Powders with a high apparent density have a high green density, and sintering-induced densification is easy. Specifically, the addition of only a small amount of sintering aid is sufficient for dense sintering.

REFERENCES

1. Abe, O., Saito, K., Kanzaki, S., Tabata, H., Kume, S. & Machida, M., Effect of Powder Characteristics on Sinterability of Silicon Nitride Ceramics. *Reports of the Government Industrial Research Institute,* Nagoya, **34** (1985), 355–66.
2. Wötting, G. & Ziegler, G., Powder Characteristics and Sintering Behaviour of Si_3N_4-Powders. *Powder Metal. Intern.,* **18** (1986), 25–32.
3. Inomata, Y., Stability Relationship in the System of β-Si_3N_4-α-Si_3N_4-Si_2N_2O and their Structural Change by Heating above 1600°C. *Yogyo-kyokai-shi,* **82** (1974), 522–6.
4. Matsumoto, A., Hayashi, H., Ichikawa, K. & Noda, T., Fine Silicon Nitride and Silicon Carbide Powders. *Taikabutsu,* **38** (1986), 526–8.

10 | Properties and Applications of Si_3N_4 Whiskers

K. NIWANO

ABSTRACT

The manufacturing processes and properties of silicon nitride whiskers are outlined, and their applications to aluminum metal, epoxy resin, polyamide, ceramic fiber, and plasma spray coating are described. Silicon nitride whisker reinforcement improves the properties of composite materials remarkably.

10.1 INTRODUCTION

Whiskers are also referred to as crystal whiskers, and since it was discovered that they have a strength approaching the theoretical value,[1] synthesis research has been conducted and metallic and various inorganic compound whiskers have been fabricated. However, all of these attempts have been confined to prototype efforts on a laboratory scale.

With the recent activity in composite materials development, diversified demands for reinforced fibers have been frustrated by a lack of whisker mass production techniques. Recently, industrial-scale production of silicon nitride, SiC, and potassium titanate whiskers has finally begun, and whiskers have found their way into the ranks of industrial materials.

10.2 FABRICATION AND CHARACTERISTICS OF SILICON NITRIDE WHISKERS

There are three synthesis methods for silicon nitride whiskers, just as for silicon nitride powders; (1) metallic silicon nitriding, (2) silica reduction, and (3) silicon halide. However, in order to form whiskers, further efforts are required.

A silicon nitride whisker manufacturing device that uses metallic silicon nitriding[2] is shown in Fig. 10.1. Raw material silicon is placed in the crucible, Si vapor is generated by heating, the vapor is sent from below to the reactor chamber using Ar gas as a carrier gas, and NH_3 is introduced from above as a reaction gas to form silicon nitride whiskers on the graphite baffle.

A silicon nitride whisker manufacturing device utilizing silica reduction[3] is shown in Fig. 10.2. The device consists of multi-storied shelves. A mixture of a silica-containing raw material and carbon is placed on the shelves, and N_2 gas is introduced from below. Whisker formation conditions for various raw materials, using the device shown in Fig. 10.2, are given in Table 10.1. Good results were obtained for the colloidal silica + Si + lampblack system. An industrial manufacturing furnace[4] is shown in Fig. 10.3.

Fig. 10.1. Device for fabricating silicon nitride whiskers using metallic silicon nitriding.

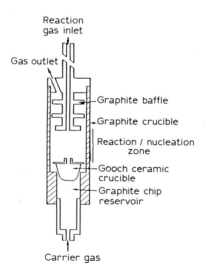

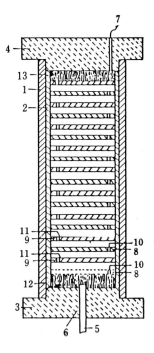

Fig. 10.2. Device for fabricating silicon nitride whiskers using silica reduction. 1, Al_2O_3 tube; 2, graphite lining; 3, 4, refractory brick plug (aluminum silicate); 5, N_2 inlet tube; 6, holes; 7, exhaust holes; 8, aluminum silicate disk; 9, graphite disk; 10, holes; 11, granular graphite.

In the silicon halide method, for example, silicon tetrahalide and ammonia are reacted in the liquid or vapor phase to form silicon diimide, which is then heat-treated to obtain amorphous silicon nitride powder. A method is known in which this amorphous silicon nitride is then blended with Fe (or an Fe compound) and silicon dioxide (or a silicon dioxide-containing compound) and baked in a nonoxidizing gas atmosphere.[5]

Detailed research has been conducted on the Si whisker growth mechanism, and the vapor–liquid–solid (VLS) mechanism using impurities as parameters has been proposed.[6]

The VLS mechanism is also believed to be at work in silicon nitride whiskers, with impurities playing an essential role in whisker growth. The above-described Fe compound is one such example.

The characteristics of silicon nitride whiskers manufactured by Tateho Chemical Industries are shown in Table 10.2. α-Si_3N_4 is the main crystal phase, with β-Si_3N_4 present in small quantities.

Table 10.1 Various Raw Materials and Si$_3$N$_4$ Whisker Formation

Disk no.	Raw material mixture	Resulting Si$_3$N$_4$ whisker growth on raw materials
1 and 2	87% Al$_2$Si$_2$O$_7$, 13% lampblack	Good
3	60 mesh SiC	Fair
4	87% diatomaceous earth, 13% lampblack	Fair
6	87% wollastonite, 13% lampblack	Fair
7	87% Al$_2$Si$_2$O$_7$, 13% lampblack	Good
8	87% synthetic magnesium silicate, 13% lampblack	Fair
9	87% pyrophyllite, 13% lampblack	Fair
10	87% kaolin, 13% lampblack	Good
11	87% feldspar, 13% lampblack	Fair
12 and 13	58% colloidal SiO$_2$, 27% Si metal, 15% lampblack	Very good
14	57% silicic acid, 29% Si metal, 14% lampblack; small amount of Al$_2$O$_3$, cement binder	Good
15 and 16	19% colloidal Al$_2$O$_3$, 66% colloidal SiO$_2$, 15% lampblack	Fair
17 and 18	85% colloidal SiO$_2$, 15% lampblack	Good
21 and 22	87% Al$_2$Si$_2$O$_7$, 13% lampblack, gum arabic binder	Good
23 and 24	95% Al$_2$Si$_2$O$_7$, 5% lampblack, gum arabic binder	Good
25 and 26	83% Al$_2$Si$_2$O$_7$, 17% lampblack, gum arabic binder	Good

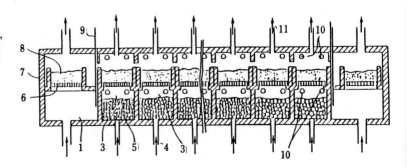

Fig. 10.3. Furnace for industrial fabrication of silicon nitride whiskers. 1, Inlet gas substitution chamber; 2, outlet gas substitution chamber; 3, small chamber; 4, gas blowing port; 5, carbon pellets; 6, fine holes; 7, tray; 8, raw material; 9, sliding door; 10, resistance heating element; 11, gas outlet.

Table 10.2
Characteristics of Silicon Nitride Whiskers Produced by Tateho Chemical Industries

Product name	SNW
Chemical formula	Si_3N_4
Crystalline phase	α type
Diameter	0·1–1·6 μm
Length	5–200 μm
Density	3·18 kg/m³
Chemical composition (%)	
Mg	0·2 (max.)
Ca	0·5 (max.)
Al	0·2 (max.)
Fe	0·1 (max.)
Moh's hardness	9
Thermal expansion coefficient	2·5 (10^{-6}/°C)
Electrical properties	Insulator
Chemical properties	Corroded by NaOH and HF but stable with respect to most acids

The length and diameter distribution are shown in Figs 10.4 and 10.5. The mean length is approximately 50 μm, and the mean diameter is approximately 0·9 μm: Fig. 10.6 is a scanning electron micrograph (SEM). According to observation by transmission electron microscopy (TEM),[7] two preferential growth orientations have been found: an orientation perpendicular to the $\{10\bar{1}0\}$ plane and an orientation perpendicular to the $\{10\bar{1}1\}$ plane. Whiskers growing with the former orientation were found to be triangular prisms, with necks observed occasionally. Branches developed, with these necks as origins. Whiskers

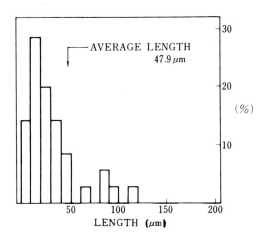

Fig. 10.4. Whisker length distribution.

Fig. 10.5. Whisker diameter distribution.

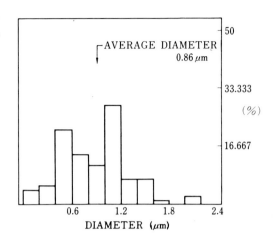

growing with an orientation perpendicular to the $\{10\bar{1}1\}$ plane, on the other hand, were tape-shaped, and it has been reported that planar defects consisting of the $\{0001\}$ and $\{10\bar{1}2\}$ planar defects are spread at high density, concentrating on one side of the whiskers.

Fig. 10.6. SEM photomicrograph of silicon nitride whiskers.

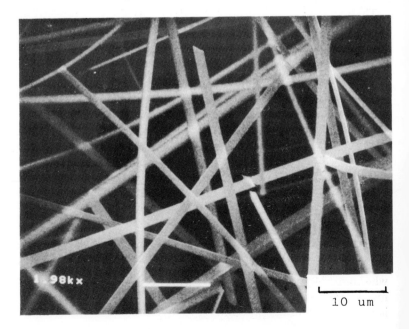

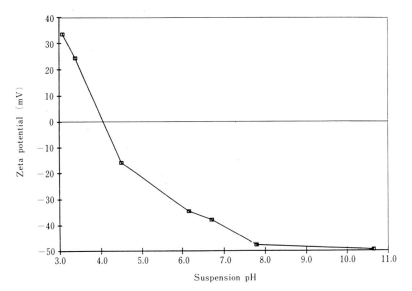

Fig. 10.7. Zeta potential of silicon nitride whiskers.

Concerning whisker dispersion, Fig. 10.7 shows zeta potential measurements.[8] An isoelectric point of 4·1 was obtained.

The measurement of silicon nitride whisker tensile strength is difficult, but it is reported[9] that, in the range 60–1500 kgf/mm², silicon nitride whiskers exhibit dimensional dependence shown by

$$\sigma_t = \text{constant} \times A_s^{-\alpha_3}$$

where σ_t is the tensile strength, A_s is the surface area, and $0{\cdot}8 < \alpha_3 < 1{\cdot}1$.

10.3 APPLICATIONS OF SILICON NITRIDE WHISKERS

10.3.1 Whisker-reinforced Metals (WRM)

As shown in Fig. 10.8, there are two main methods of preparing whisker-reinforced metals: squeeze casting (a) and powder metallurgy (b). In the former method, whiskers are formed in advance as preforms, and molten metal is infiltrated under

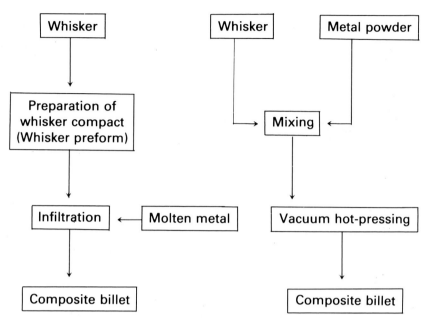

Fig. 10.8. Methods of fabricating whisker–metal composites.

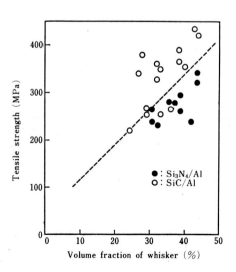

Fig. 10.9. Tensile strength of whisker-reinforced aluminum composites.

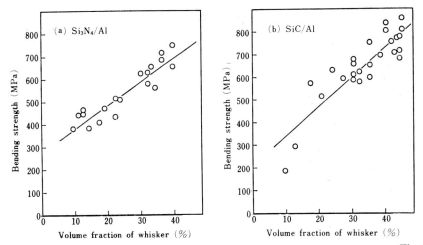

Fig. 10.10. Bending strength of whisker-reinforced aluminum composites.

pressure. In the latter method, whiskers are mixed with a metal powder and a composite is formed with vacuum hot-pressing.

The mechanical characteristics of whisker-reinforced pure aluminum[10] are shown in Figs 10.9–10.11. Although SiC whiskers offer a slightly higher tensile strength than silicon nitride whiskers, the bending strength and elastic modulus were virtually identical for both materials. The results of hardness measurements are shown in Table 10.3;[10] remarkable increases were effected by both Si_3N_4 and SiC.

Fig. 10.11. Elastic modulus of whisker-reinforced aluminum composites.

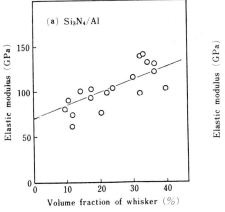
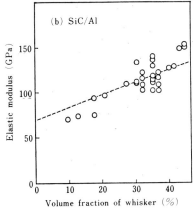

Table 10.3
Vickers Hardness of Whisker–Al Composites

Composites	$V_w = 20\%$	$V_w = 30\%$	$V_w = 40\%$
Si_3N_4–Al	130	190	250
SiC–Al	130	180	240

Whisker-reinforced metals have an advantage over long-fiber-reinforced metals in that they can be processed using conventional secondary-metal processing techniques such as extrusion and rolling.

10.3.2 Whisker-reinforced Plastics (WRP)

Methods of manufacturing whisker-reinforced plastics differ with the type of plastic, but the same method can be applied as to various powder fillers.

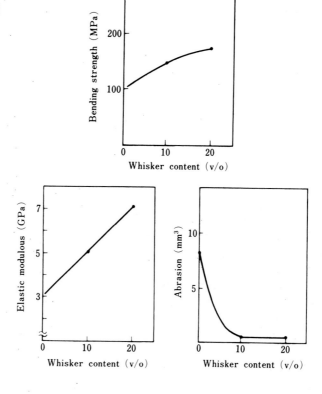

Fig. 10.12. Characteristics of silicon nitride whisker–epoxy resin composites.

Table 10.4
Characteristics of Silicon Nitride Whisker–Polyamide Resin Composites

Whisker weight (%)	PA 66 CM 3001 N matrix		
	0	20	40
Tensile strength (MPa)	79	74	103
Elongation (%)	110[a]	2·8	3·8
Bending strength (MPa)	111	136	177
Elastic modulus (GPa)	2·6	3·9	6·6
Rockwell hardness (R scale)	108	114	117
IZOD impact value (notched) (J/m)	75	45	46
Heat deformation temperatures (1·82 MPa)	76	158	204
Density ($\times 10^3$ g/m^3)	1·14	1·34	1·55
Ash (%)			38·7

[a] Toray CM 3001 N catalog value.

An example of a whisker–epoxy resin composite is shown in Fig. 10.12 (Technical data of the Tateho Chemical Industries Co., Ltd). The result is a remarkable improvement in elastic modulus and wear resistance. The characteristics of an injection-molded polyamide, using a composite pellet as an example of a thermoplastic resin, (Technical data of Tateho Chemical Industries Co., Ltd.) are shown in Table 10.4. A remarkable increase in the thermal deformation temperature can be noted.

10.3.3 Whisker-reinforced Ceramics (WRC)

Ordinary ceramic manufacturing techniques can also be applied to the manufacture of whisker-reinforced ceramics, but some additional effort is required in the dispersion of whiskers.

It has recently been reported that SiC whiskers contribute to the improvement of fracture toughness,[11,12] and active research is now being conducted. However, no such reports have been made concerning silicon nitride whiskers. A ceramic fiber–silicon nitride whisker composite is shown in Figs 10.13 and 10.14 (Technical data of the Isolite Insulating Products Co., Ltd). It can be seen that the addition of silicon nitride whiskers contributes to the prevention of ceramic fiber contraction and (at 50 vol%) allows for a 3- to 3·5-fold increase in strength. It has also been reported (Technical data of Isolite Insulating Products

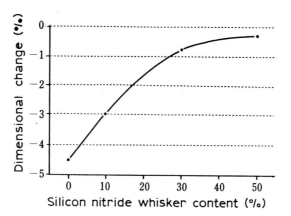

Fig. 10.13. Relation between silicon nitride whisker content and dimensional change when silicon nitride whiskers are mixed into ceramic fiber (at 1300°C for 3 h).

Co., Ltd.), in molten aluminum dipping tests, that the resulting composite material shows excellent corrosion resistance with almost no corrosion.

The results of plasma flame spraying film tests on a whisker–alumina composite (Technical data of Showa Denko K.K.) are shown in Figs 10.15 and 10.16. Major improvements

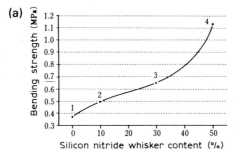

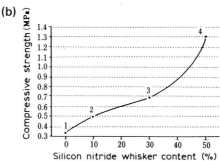

Fig. 10.14. Silicon nitride whisker content versus (a) bending strength and (b) compressive strength when silicon nitride whiskers are mixed into ceramic fiber (at 1300°C for 3 h). Bulk densities are (1) 0·47, (2) 0·46, (3) 0·41, and (4) 0·44.

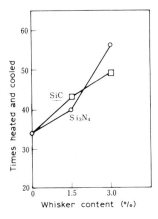

Fig. 10.15. Thermal shock resistance of Al₂O₃–whisker composite plasma flame-spraying films (heated at 1100°C for 45 s, then cooled for 45 s).

have also been shown in impact resistance, adhesive strength, and thermal shock resistance.

10.4 CONCLUSIONS

Methods of fabricating silicon nitride whiskers and their properties and applications have been outlined. With a few exceptions, whisker-reinforced composites are still in the de-

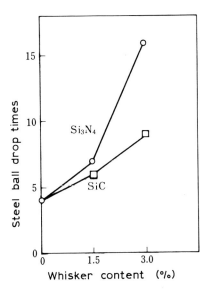

Fig. 10.16. Peeling resistance of Al₂O₃–whisker composite flame-spraying films (weight of steel ball = 110 g; dropped from a height of 82·5 cm).

velopment stage. However, active development is being conducted with the aim of applying these materials to automobile and airplane components, and to sports and leisure goods.

It has recently been confirmed that silicon carbide whiskers can contribute to the improvement of fracture toughness, particularly in ceramic materials. In addition, SiC whisker-reinforced alumina composites are being commercialized for use as cutting tools.

As our understanding of whisker characteristics grows, the establishment of composite technologies utilizing these characteristics can be expected.

REFERENCES

1. Herring, C. & Galt, J. K., Elastic properties of very small metal specimens. *Phys. Rev.*, **85** (1952), 1060.
2. Cunningham, A. L. & Davis, L. G., Preparation and characterization of a novel form of Si_3N_4 fiber. *SAMPE*, **15** (1969), 120.
3. Johnson, R. C., Alley, John K., Warwick, Wilbur H., Shell, Haskiel, R., Synthesis of fibrous silicon nitride. U.S. Patent 3,244,480 (1966).
4. Tanaka, M. & Kawabe, T., Method of manufacturing silicon nitride whisker. Japanese Patent 1,324,479 (1986).
5. Kohtoku, Y. & Masunaga, K., Production of silicon nitride whisker. Japanese Patent Provisional Publication 61-275199 (1986).
6. Wagner, R. S. & Ellis, W. C., The vapour–liquid–solid mechanism of crystal growth and its applications to silicon. *Trans. Mat. Soc., AIME*, **233** (1965), 1053.
7. Sasaki, K., Kuroda, K., Imura, T. & Saka, H., Microstructural investigation of α-Si_3N_4 whisker by transmission electron microscopy. *Yogyo-Kyokai-Shi*, **94**(8) (1986), 773–8.
8. Krishnan, S. V., Pascucci, M. R., Adai, J. H. & Mutsuddy, B. C., Interfacial reactions of organic molecules on silicon carbide and silicon nitride whiskers in aqueous and non-aqueous suspension. Presented at Am. Ceram. Soc. Annual Meeting, 1985.
9. Bayer, P. D. & Cooper, R. E., Size strength effects in sapphire and silicon nitride whiskers at 20°C. *J. Mat. Sci.*, **2** (1967) 233–7.
10. Fukunaga, H., Goda, K. & Tabata, N., Preparation and mechanical properties of whisker reinforced aluminum composites. *J. Soc. Mat. Sci. Japan*, **34**(376) (1985), 64–9.
11. Wei, G. C. & Becher, P. F., Development of SiC-whisker-reinforced ceramics. *Am. Ceram. Soc. Bull.*, **64**(2) (1985), 298–304.
12. Kandori, T., Homma, T. & Wada, S., SiC whisker reinforced Si_3N_4 sintered bodies. Yogyo-Kyokai 1986 Annual Meeting, Extended Abstracts, pp. 133–4.

11 | Joining of Si_3N_4

N. IWAMOTO

ABSTRACT

Progress in studies on Si_3N_4–Si_3N_4 and Si_3N_4–metal-or-alloy bonding is summarized. Studies on the reaction between various metals and Si_3N_4 are also given, and the compatibility behavior is reviewed. Because of the importance to silicide formation at the interface, fundamental problems on solid solubility between various metals and silicon formed from the dissociation of Si_3N_4, the diffusion behavior of metals in silicon, and the formation energy of silicide, etc., are given. Furthermore, the effect of the nickel and cobalt content on the phase stability of austenitic steels during joining with Si_3N_4 is discussed. Finally, progress in oxide solder development for the joining of Si_3N_4 with Si_3N_4 is reviewed.

11.1 INTRODUCTION

Due to its excellent high-temperature strength and resistance to thermal shock, fatigue, and mechanical detachment and corrosion, Si_3N_4, together with the sialons, is indispensable for application to diesel, Stirling, and gas turbine engines.

In particular, strategic considerations, such as increased turbine inlet temperature, energy conservation, high corrosion resistance (during the use of low-quality fuels), and the resulting possibility of operation at higher temperatures; the reduction of heavy oil consumption; and the reduced need to secure precious metals such as Ni, Cr, Co, and Nb have caused a great deal of attention to be focused on Si_3N_4.[1]

Reports concerning the basic characteristics of Si_3N_4 have been given at numerous international symposia.[1-6]

11.2 TRENDS IN JOINING RESEARCH

In Japan, a great deal of research concerning the joining of Si_3N_4 with Si_3N_4, or with different materials, as well as the reaction of Si_3N_4 with metals and alloys, has been presented at the Japan

Table 11.1 Oral Presentations at the Meetings of the Japan Institute of Metals

	April 1985		October 1985	
Substance	[SiC]	(4)	[SiC]	(5)
Thema	Reaction Insert	$\binom{3}{1}$		$\binom{3}{2}$
Species	Al, Al–Si, Fe–C, Cu–&Ni–solder, CaO–Al$_2$O$_3$–TiO$_2$		Al, Ni, Ni–Y, Al–Si–Cu, Cu–Mn + C, SiC, SiC + Ni, NaF–LiF	
Substance	[Si$_3$N$_4$]	(6)	[Si$_3$N$_4$]	(21)
Thema	Reaction Insert	$\binom{1}{5}$		$\binom{10}{12}$
Species	Al, Ti, Fe, Co, Ni, Zr, Ag–Cu–Ti		Al, Cu, Ti, Cr, Cu–Ti, Ni–Ti, Ni–Cr, Fe–Ni–Cr, Koval, Al–Ti–V, Ag–CuO + Si$_3$N$_4$	
Substance	[Sialon]	(1)	[Sialon]	(1)
Thema	Insert	1		1
Species	Al–Si		Cu–Mn + C	
Substance	[Al$_2$O$_3$]	(1)	[Al$_2$O$_3$]	(2)
Thema	Pressure bonding	1	Insert	2
Species	Ni		Fe–FeO, Al–Si	
Substance	[ZrO$_2$]	0	[ZrO$_2$]	(2)
Thema			Reaction	2
Species			Al, Ni, Al–Mg	
Substance	[MgO]	0	[MgO]	(1)
Thema			Pressure bonding	1
Species			Ni	

Table 11.1—*contd.*

	April 1986		October 1986	
Substance	[SiC]	(4)	[SiC]	(④)
Thema	Joining Reaction	$\binom{2}{2}$	Wetting Joining Reaction	$\binom{1}{2}{1}$
Species	Ni–ZrH$_2$, Ni, Fe, Cr, Ni–Cr		Ti–Cu, Zr–Cu, Ni + Ti, Cu, Ni–Cr, Ni–Mo, Ni–Ti, Ni–W, Ni–Nb	
Substance	[Si$_3$N$_4$]	(⑭)	[Si$_3$N$_4$]	(⑱)
Thema	Reaction TEM Joining	$\binom{2}{1}{11}$	Reaction TEM Joining	$\binom{3}{2}{13}$
Species	Al, Ni–Al, Ni–Al$_2$O$_3$, Ni, Ni–W, Ni–W–Cu, 6061, 2017, BaO–SiO$_2$, SrO–SiO$_2$, M$_x$O$_y$–SiO$_2$, Al–Cu, Au–Cu–Ti, Ni–Cr, Al$_2$O$_3$, Ni, BAg8, Ti, Mo–Mn, CuS		Al, Ni + Ti, Al–Ni, Cu, Cu + S$_{45}$C, Ni–Al$_2$O$_3$, Cu, Ni, Ti, Mo, W, La$_2$O$_3$–Y$_2$O$_3$–Al$_2$O$_3$–SiO$_2$–BeO, Ca–Si–Y–Al–O, Mo–Si, Fe, Ag–Cu–Ti, Al–Si	
Substance	[Al$_2$O$_3$]	(③)	[Al$_2$O$_3$]	(②)
Thema	Joining Wetting	$\binom{2}{1}$	Joining TEM	$\binom{1}{1}$
Species	Al–alloy, Al–Si–Mg, Al–Si, Fe$_2$O$_3$, Si$_3$N$_4$		Al, Al–Si	
Substance	[ZrO$_2$]	(③)	[ZrO$_2$]	(⑤)
Thema	Joining	(3)	Joining Reaction	$\binom{4}{1}$
Species	Mo–Mn–MnO–SiO$_2$–Al$_2$O$_3$, Pt, Cu–Ti		Ti–Cu, Zr–Cu, steel, Pt, MnO–SiO–Al$_2$O$_3$ + Mo, La$_2$O$_3$–Y$_2$O$_3$–Al$_2$O$_3$–SiO$_2$–BeO	
Substance	[MgO]	(③)	[MgO]	(②)
Thema	Joining	(3)	Joining	(2)
Species	Cu–Ti, Fe, Fe$_2$O$_3$		Ni, Co	
Substance			[AlN]	(①)
Thema			Joining	(1)
Species			Ti–Cu, Zr–Cu	

Institute of Metals' spring and fall assemblies durng the past two years: details are listed in Table 11.1. The number of different materials on which tests are being conducted is surprising, and the attention being given to these topics can be understood.

Methods announced up to now can be broadly grouped into

the following categories:

(1) Si_3N_4 and various additives are mixed on the interface.[7-9]
(2) ZrO_2 or $ZrSiO_4$ powder is used on the interface.[10]
(3) A metal is used.[11-13]
(4) Si_3N_4 powder is used on the interface, and the material is hot-pressed.[14]
(5) Nothing is used on the interface, and the materials are directly joined.[15]
(6) The materials are bonded without the application of pressure.[15]
(7) Oxides and metals are reacted.[16]
(8) Si-containing alloy solder is used.[17]
(9) Oxide glass is used.[18-24]

In (1), when the mixture contains Y_2O_3 and the reaction occurs at 1800°C under an applied pressure of 3 GPa, it has been reported that Y migrates into the bulk, and bulk hardness drops at temperatures above 1000°C.[7] In addition, in tests supplying nitrogen during the reaction with Si to secondarily form Si_3N_4 on the interface, strength degradation has resulted from the generation of pores, and this approach is consequently unsuited to applications requiring high strength.

In the case of (2), doping with ZrO_2 is effective, and with the use of hot-pressed Si_3N_4 a joining strength of 175 MPa was obtained, even with a low applied pressure (e.g., <1·5 MPa). It has been reported, however, that the reaction is weak when $ZrSiO_4$ is used.[10]

In the case of (3), it is possible to form, on the joining interface, a substance of the type that exists on Si_3N_4 grain boundaries, by using an oxide that is effective in Si_3N_4 sintering; but high-temperature strength degradation is believed to occur due to formation of a large quantity of second phases.

Those metals which can be coupled to joining materials or bonded directly to Si_3N_4 are: Zr, Al, Cu, Si, Si alloys, and Mo.

The disadvantages of hot-pressing are the long time required and the impossibility of application to materials with complicated shapes. Therefore the use of oxide glass has been proposed as a technique allowing application to complicated shapes, physical and chemical compatibility with Si_3N_4, selection over a wide temperature range, and free setting of the useful temperature.[18-24] These results will be described below.

11.3 REACTION OF Si_3N_4 AND VARIOUS METALS

11.3.1 Aluminum, Nickel

The high-temperature reaction of Si_3N_4 whiskers and Al or Ni vapor deposition layers was investigated in 1966.[25] In the reaction with Al, Si_3N_4 decomposition was observed upon heating at 630°C for 30 min in flowing Ar. In the reaction with Ni, the formation of Ni-silicide was assumed for heating at 1000°C for 15 min. However, the results of this experiment did not predict a reaction with bulk Si_3N_4.

In 1972 differences in the reaction of Ni-covered Si_3N_4 whiskers in vacuum and in Ar and N_2 atmospheres were studied.[26] In vacuum, Si_3N_4 decomposed at 1100°C. In Ar, crystal growth from the whisker sidewalls occurred, but the formation of Ni-silicide was not observed. In the N_2 atmosphere no change was observed:

$$Si_3N_4 + 3Ni \rightarrow 3NiSi + 2N_2 \quad \text{(in vacuum)}$$

The equilibrium nitrogen partial pressure of Si_3N_4 at 1000°C is calculated to be $P_{N_2} \sim 2 \cdot 5 \times 10^{-5}$ atm, with decomposition occurring only in vacuum (10^{-7}–10^{-8} atm):

$$Si_3N_4 \rightarrow 3Si + 2N_2$$

In processing in an Ar atmosphere, observed crystal growth from the sidewalls is seen as SiO_2 growth resulting from the catalytic effect of Ni.

11.3.2 Tin, Tin–Titanium

The relation of the Si_3N_4–molten Sn contact angle to temperature and atmosphere is shown in Table 11.2.[27] According to these results, in a H_2 atmosphere the contact angle does not change, even when temperature is increased. In vacuum, however, it can be seen that the contact angle drops drastically when the temperature reaches 1200°C due to the decomposition of SiO_2 formed on the surface.

The amount of Ti added to the Sn–Ti alloy vs. the interface

Table 11.2
Contact Angle of Liquid Tin[27]

System		800°C	1100°C	1200°C	1325°C	1470°C
Al_2O_3	H_2	161	174	165	—	149
	In vacuo	162	167	166	—	—
SiC	H_2	166	165	160	—	140
	In vacuo	162	165	75	—	—
Si_3N_4	H_2	154	154	—	—	140
	In vacuo	168	154	29	—	—
$MoSi_2$	H_2	166	154	—	53	28
	In vacuo	170	35	20	—	—

energy is shown in Fig. 11.1. For the reason described above, SiO_2 formed on the surface during processing at high temperatures and in vacuum decomposes, and when sizeable amounts of Ti are contained in the alloy, Ti is adsorbed in the interface and the interface energy is reduced.

11.3.3 Copper

At 1250°C surface tension causes small drops to form on the hot-pressed silicon nitride. A method of reducing surface tension

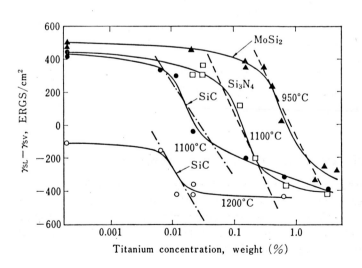

Fig. 11.1. Difference between interface and solid-surface energy for the systems Sn–Ti (1)–$MoSi_2$, SiC, and Si_3N_4(s).[27]

by treating the hot-pressed silicon nitride surface with a $KMnO_4 + CuSO_4$ aqueous solution at 60°C for 10 min and then performing surface oxidation has been suggested. In this case, joining was observed after 100-μm Cu foil was placed on the surface and heated to 1075°C in N_2. Elemental distribution analysis using XMA showed no penetration of Cu into the Si_3N_4. In addition, homogeneous dispersion of Cu did not occur, and the generation of countless pores and perpendicular cracks was observed.[28]

11.3.4 Aluminum

The pre-reduction of Al is required for dense bonding of Al–Si_3N_4. For example, even when an Al foil was placed on the surface and treated at 1050–1250°C in a C-containing gas or in vacuum, no reaction occurred at 1050–1150°C in vacuum, and disappearance of the Al foil was observed when the temperature reached 1250°C. In the case of hot-pressed silicon nitride this signifies the beginning of decomposition, as well as the beginning of a strong decomposition of the RBSN.

The generation of parallel pores in the interphase was observed, and XMA analysis showed high-concentration Al migration to the intermediate layer, 200-μm Al migration to the Si_3N_4 side (of the Al layer) in the case of RBSN, and Si migration to the intermediate layer.[28]

Al was sandwiched between Si_3N_4 layers and heat-treated at 1470–1970 K under a pressure of 0–20 MPa in 1-atm N_2. Under conditions of 1970 K at 20 MPa, a bending strength of 140 MPa was obtained at 1670 K. The formation of AlN-system sialons on the interface has been proposed as the reason for this, suggesting a dependence on pressure.[29]

11.3.5 Tantalum, Molybdenum, Tungsten

An intermediate layer was formed at 1400–1470°C, but in processing in vacuum and in N_2 neither joining nor reaction with Si_3N_4 was observed. Decomposition of the Si_3N_4 surface began at 1400–1470°C in vacuum, and in N_2 in particular Ta exhibited discoloration. Mo and W lost their metallic luster.[28]

11.3.6 Nickel–30 w/t Molybdenum

The contact angle was 0° at reduced pressure (6×10^{-4} Torr). It is believed that Si_3N_4 decomposition and the precipitation of Si facilitated wetting.[30]

11.3.7 Silicon

Numerous workers have reported that Si_3N_4 shows wetting.[31–33]

Concerning the wetting of molten Si on a CVD Si_3N_4 film, the results of an investigation into the influence of P_{O_2} are shown in Fig. 11.2. It can be seen that the contact angle increases with P_{O_2}. Although an Si–SiO_2 equilibrium has been established on the interface, when P_{O_2} is high SiO_2 is formed and evaporated, resulting in a cleaning of the molten Si surface and an increase in interface energy.[34]

11.3.8 Zirconium

A method in which Zr is sandwiched between layers of Si_3N_4 and Zr, or Si_3N_4 and Nb, has also been investigated. When the Si_3N_4 is of high purity, a high joining strength can be obtained at 1400°C, but with low-purity Si_3N_4 the maximum joining strength

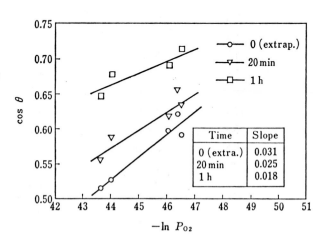

Fig. 11.2. Relation between $\cos\theta$ and $\ln P_{O_2}$ for CNTD Si_3N_4 at various times.[37]

Table 11.3 Compatibility of Metals and Si_3N_4 (in vacuum or inert gas)[38]

Metal	Temperature (°C)	Time	Reaction	Ref.
Al	800	950 h	○	40
	900	200 h	○	39
	1000	300 h	○	41
Fe	1560	1 h	×	41
Cast iron	1450	2 h	×	42
Ni	1460	5 min	×	41
Nichrom	?	?	×	43
Nickel silver	?	?	○	43
Ti	1460	5 min	×	41
Cu	1150	7 h	×	40
	1200	300 h	○	41
Bronze	950	72 h	○	42
Zn	550	500 h	○	40
	940	168 h	○	41
Sn	300	144 H	○	40
	800	10 h	○	41
Pb	400	144 h	○	40
	800	10 h	○	41
Mg	750	20 h	×	40
	800	2 h	×	41
Bi	800	10 h	○	41
Cd	550	10 h	○	41
Si	1450	5 min	×	41
V	1730	5 min	×	41
Cr	1830	5 min	×	41
Mn	1270	5 min	×	41
Co	1500	5 min	×	41
Zr	1930	5 min	×	41
Ag	?	?	○	43
Au	?	?	○	43

appears to be obtained at low temperatures. Since toughness depends on Nb thickness, the use of thick Nb means that breakdown will occur within the Si_3N_4.[35]

11.3.9 Tungsten, Molybdenum

When Si_3N_4 is to be joined with W or Mo, a method in which Si_3N_4 + W or Si_3N_4 + Mo is inserted in the intermediate layer has been suggested.[36]

11.3.10 Copper, Molybdenum

When Si_3N_4 is joined with Cu or Mo, AgCuTi is used as a soldering material. The metallizing temperature on the Cu is set at 820°C.[37] Work done by Hirai & Matsuda[38] concerning the reaction of Si_3N_4 and various molten metals in vacuum or inert gas is summarized in Table 11.3.[39–43] The ease or difficulty of silicide formation from the standard free energy of each reaction is shown in Fig. 11.3.

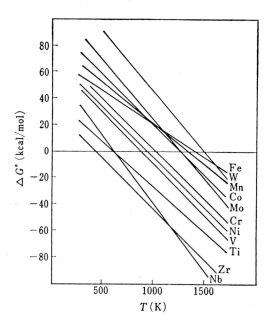

Fig. 11.3. A standard free energy diagram for nitrides.[38] $Si_3N_4 + M \rightarrow$ M-silicide + $2N_2$.

$Si_3N_4 + 9Fe \rightarrow 3Fe_3Si + 2N_2$
$Si_3N_4 + 5W \rightarrow W_5Si_3 + 2N_2$
$Si_3N_4 + 9Mn \rightarrow 3Mn_3Si + 2N_2$
$Si_3N_4 + 6Co \rightarrow 3Co_2Si + 2N_2$
$Si_3N_4 + 5Mo \rightarrow Mo_5Si_3 + 2N_2$
$Si_3N_4 + 9Cr \rightarrow 3Cr_3Si + 2N_2$
$Si_3N_4 + 9Ni \rightarrow 3Ni_3Si + 2N_2$
$Si_3N_4 + 3/2V \rightarrow 3/2VSi_2 + 2N_2$
$Si_3N_4 + 5Ti \rightarrow Ti_5Si_3 + 2N_2$
$Si_3N_4 + 5Zr \rightarrow Zr_5Si_3 + 2N_2$
$Si_3N_4 + 5Nb \rightarrow Nb_5Si_2 + 2N_2$

11.4 COMPATIBILITY OF Si_3N_4 AND ALLOYS

In various research conducted outside Japan, the object of study has not been the direct joining of alloys and Si_3N_4 but rather the compatibility of the two when an alloy and Si_3N_4 make contact and are brought to a high temperature.

In 1976, Si_3N_4 and IN-718 were maintained in contact at 1150°C in an air atmosphere. The reaction was dominated by the SiO_2 layer formed on the Si_3N_4 surface, and TiO_2 and NbO were formed from trace amounts of Ti and Nb contained in the steel according to the following reaction:

$$Ti(s) + SiO_2(s) = TiO_2(s) + Si(s)$$

As a result, the formation of a precipitate having high Ti and Nb contents was reported.[44]

In 1979 the contact compatibility of Fe-, Ni-, and Mo-matrix alloys and Si_3N_4 was investigated.[45] The test was carried out at 800–1100°C for 1010–5161 h.

The penetration of Fe, Cr, and Ni into the Si_3N_4 surface and the depth of corrosion for Si_3N_4 and the alloys are shown in Table 11.4:

(1) In the Fe–Ni–C and Fe–Cr systems, an adhesive silicide

$$Cr(Fe/Ni) + Si_3N_4 \rightarrow Cr(Fe/Ni)(Si,N)$$

was formed.

(2) In those alloys containing Mo, Si and nitrogen resulting from Si_3N_4 decomposition formed Mo-silicides; in those alloys containing Ti, Al, or Y, nitrides of these elements were formed.

In the same year the reaction of Ni-matrix alloys and Si_3N_4 at 700–1150°C under applied pressure and low P_{O_2} conditions was investigated, but it was reported that the reaction did not show significant progress.[46]

11.5 THE FOUNDATION FOR SILICIDE FORMATION

11.5.1 Solubility of Si in Various Metals

As shown in Fig. 11.4, it is possible to dissolve large quantities of Si (e.g., 10 at% Si) in Ni, Co, Fe, and V, having an ion radius

Table 11.4 Reaction of Si_3N_4 with Stainless Steel and Alloy[45]

Stainless steel and alloy	Exposure conditions		Metallic pick-up on Si_3N_4 surface (µg/cm^2)			Maximum depth of attack (µm)	
	Temperature (°C)	Time (h)	Iron	Chromium	Nickel	Si_3N_4	Steel
Fecralloy[a]	800	4078	0	1·5–36		NM	NM
	825	1010	0	2		NM	NM
		5009	0	8, 15		NM	NM
	900	1010	0	28		NM	NM (63)[b]
		5161	0	100, 155		1	2 (75)
	1000[a]	1173	10, 40	68, 105		2	7 (0)
		1173	110, 175	290, 395		3	4 (16)
		4030	ND	ND		20	6 (80)
	1100	1125	245–270	370–525		21	17 (375)
316	800	4824	0	0·3, 1·3	0	NM	NM
	900	1172	0	30, 38	2, 3	NM	5
	1000	1075	80	130	14	39	82
20/25/Nb	800	4078	0	0·2–0·7	0	NM	NM
	825	1010	0	1·0	0	NM	NM
		5009	0	4, 10	0	NM	NM
	900	1010	0	7·5	0	NM	NM
		5161	0	110, 150	4, 6	~2	
	1000	1173	15, 170	570, 580	430, 485	47–150	110–175
PE16	825	1010	0	1·9	0	NM	NM
		5009	0	12, 21	0·5, 1·0	NM	NM
	900	1010	0	28	0	NM	NM
		5161	0	47, 75	0–0·8	NM	1
	1000	1173	42, 65	265, 355	285, 340	6	23
Hastelloy X	800	4078	0	0·4–2·3	0	NM	NM
		4824	0	0·3, 1·3	0	NM	NM
	900	1172	0	46, 260	1·2, 14	4	3
	1000	1173	80, 125	500	585, 820	16	56
		4644	ND	ND	ND	65	65
Nimonic 75	800	4078		2·0–3·5	0	NM	NM
		4824		0·3–0·7	0	NM	NM
	900	1172		38, 70	11, 12	3	NM
	1000	1173		550	935, 1000	10	85
		4644		ND	ND	80	95

NM, none measurable; ND, not determined, gross attack.
[a] Results are for the yttrium-bearing Fecralloy steel except that indicated at 1000°C, which is for the yttrium-free material.
[b] Values in parentheses are depths of intergranular precipitates.

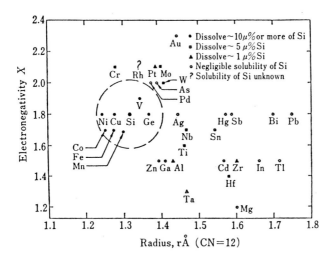

Fig. 11.4. Darken and Gurry plot of electronegativity vs. atomic radius to show the solubility of Si in various metals.

close to that of Si. However, elements with larger atomic radius than Si, are virtually insoluble in Si.[47]

11.5.2 Diffusion of Metals in Si

As shown in Fig. 11.5,[48] Cu appears to exhibit the fastest diffusion, with Ni and Fe alloys showing high values. All of the elements show larger self-diffusion coefficients than Si and Ge.[49–60] Self-diffusion and solute diffusion coefficients in Si[49,62–70] are shown in Table 11.5.[61]

11.5.3 Silicide Formation Energy

A comparison of the formation energies for various silicides[61] is shown in Table 11.6. It can be seen that for Zr and Ti, MSi is more stable than MSi_2. This is because, crystallographically, MSi has a simpler crystal structure than MSi_2. It should be noted that M_5Si_3 is formed only when the amount of Si is small.

Fig. 11.5. Survey of the diffusivities of foreign atoms in silicon.[48,49,60]

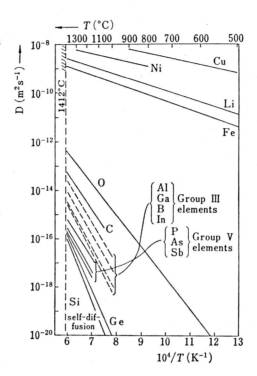

Table 11.5 Self-diffusion and Solute Diffusion; Coefficient in Si[61]

Element	$D(cm^2/s)$		Refs
Cu	$4.7 \times 10^{-3} \exp(-0.43/kT)$	(300–700°C)	49
Ag	$2 \times 10^{-3} \exp(-1.6/kT)$	(1100–1350°C)	62
Au	$2.4 \times 10^{-4} \exp(-0.39/kT)$	(700–1300°C)	63
Al	$8 \exp(-3.47/kT)$	(1085–1375°C)	64
Si	$1.81 \times 10^4 \exp(-4.86/kT)$	(900–1300°C)	65, 63
As	$60 \exp(-4.2/kT)$	(850–1150°C)	66
Cr	$<10^{-8}$	(1200°C)	67
Mn	$>2 \times 10^{-7}$	(1200°C)	68
Fe	$>5 \times 10^{-6}$	(100–1115°C)	67
Ni	1.57×10^{-7}	(800°C)	69
Pt	Similar to Au in Si		70

Table 11.6 Free Energy of Formation, ΔH, of Silicides[61]

Silicide	ΔH (kcal/g-atom)	Silicide	6-H (kcal/g-atom)	Silicide	6-H (kcal/g-atom)
Mg$_2$Si	6·2	Ti$_5$Si$_3$	17·3	V$_3$Si	6·5
		TiSi	15·5	V$_5$Si$_3$	11·8
FeSi	8·8	TiSi$_2$	10·7	VSi$_2$	24·3
FeSi$_2$	6·2				
		Zr$_2$Si	16·7	Nb$_5$Si$_3$	10·9
Co$_2$Si	9·2	Zr$_5$Si$_3$	18·3	NbSi$_2$	10·7
CoSi	12	ZrSi	18·5, 17·7		
CoSi$_2$	8·2	ZrSi$_2$	12·9, 11·9	Ta$_5$Si$_3$	9·5
				TaSi$_2$	8·7, 9·3
Ni$_2$Si	11·2, 10·5	HfSi			
NiSi	10·3	HfSi$_2$		Cr$_3$Si	7·5
				Cr$_5$Si$_3$	8
Pd$_2$Si	6·9			CrSi	7·5
PdSi	6·9			CrSi$_2$	7·7
				Mo$_3$Si	5·6
				Mo$_5$Si$_3$	8·5
Pt$_2$Si	6·9			MoSi$_2$	8·7, 10·5
PtSi	7·9			W$_5$Si$_3$	5
RhSi	8·1			WSi$_2$	7·3

11.5.4 Dominant Diffusing Elements During Silicide Formation

As shown in Table 11.7,[71] markers were used to investigate which of Si and the doped element was the dominant diffuser during silicide formation.[72–83] It can be seen that while the doped element dominates at low temperatures, Si is dominant at high temperatures. This is because of two factors: first, at high temperatures the breaking of covalent Si bonds by phonon energy facilitates the diffusion of Si; and second, structural defects, grain boundary defects, and point defects resulting from the high temperatures also aid in Si diffusion.[71]

Table 11.7 Dominant Diffusing Species in Metal–Silicon Systems[71]

System	Dominant diffuser	Refs	System	Dominant diffuser	Refs
TiSi$_2$, TiSi	Si	72, 73	WSi$_2$	Si	77
HfSi	Si	74	FeSi	Si	78
V$_3$Si	V	73, 75	Co$_2$Si	Co	72, 79
VSi$_2$	Si	73, 75	Ni$_2$Si	Ni	80, 81
TaSi$_2$	Si	76	Pd$_2$Si	Pd, Si	73
MoSi$_2$	Si	77	Pt$_2$Si	Pt	82, 83

11.5.5 Silicide Stability

Silicide thermal stability and various high-temperature characteristics remain unclear. Characteristics, and particularly the electrical characteristics, of silicides in electronic components have been thoroughly investigated.[84] Defect structures have also been well researched.[85]

11.6 PROBLEMS ON THE METAL SIDE

As described above, the formation of silicides on an electronic substrate proceeds in the low-temperature range 200–600°C regardless of the melting point of the accompanying metal. In this case, the accompanying metal forms the silicide through diffusion along the grain boundaries of the Si substrate.

However, during the joining of Si_3N_4 and metals, the reaction is one with the Si formed from decomposition of Si_3N_4 as the temperature rises. The amount thus formed is not significant, and in the case of special steels containing Ni or Co, only the diffusion of the dominant diffusers such as Ni or Co need be considered. Although it is impossible for the present author to describe in detail the structural changes that occur during the joining of numerous special steels, the following is offered from the standpoint of Ni and Co migration.

11.6.1 Effect of Co Doping

Compared with Ni and Mo, Co does not offer any significant improvement of properties as a doping element in alloyed steel.[86]

(1) Hardening: not significant.
(2) Carbides: there is no single system carbide of Co; Co affects only nucleus formation and growth rate.
(3) Solid solution hardening: no significant effect when doped in austenite but effective with martensite; hence useful in chromium steel.
(4) δ-ferrite: doping with Co has a stabilizing effect on austenite.
(5) Transformation temperature: no change of Ac_1 point.

The effect of doping Co into low-alloy steel, marageing steel, carbide-hardened martensite steel, and tool steel is well documented.[87]

Numerous reports on the effect of doping superalloys with Co have been made. Concerning the doping of Udimet 700 with Co, however, it has been reported only that a σ-phase was formed.[88-90]

11.6.2 Effect of Ni Doping

The doping effect of Ni in Ni–Cr, Ni–Fe–Cr, Ni-matrix creep-resistant steels, and Ni–Fe steels is well documented.[91]

The effect of the doping element on superalloy structure is summarized in Table 11.8.[92]

11.6.3 Effect of Ni and Co Diffusion

Compared to the structural change in steel that occurs during the joining of SiC and alloy steel, it is believed by the present author that the steel structural change accompanying elemental migration from the alloy side does not pose significant problems. Rather, the problem lies in the migration to the steel of Si from the Si_3N_4 side.

11.7 Si_3N_4–Si_3N_4 JOINING USING OXIDE SOLDER

Although the joining of Si_3N_4 and metals is naturally an important topic, so is the search for a simple method of joining Si_3N_4 with itself, since molding the various desired shapes of Si_3N_4 by casting and then sintering is extremely expensive.

Numerous additives, including the Y_2O_3–Al_2O_3 system, have been used in Si_3N_4 sintering.[1,6] Since these additives have been concentrated on the grain boundaries, it is widely believed that introducing oxide solder on the Si_3N_4–Si_3N_4 joining interface will cause them to react with the glass phase formed on the grain boundaries. The following is a summary of reports published to

Table 11.8
Phases Identified in Leading Superalloys[92]

Phases	Ni-base	Co-base	Fe-base
Matrix	Gamma, fcc (Ni, Cr, Co, ...)	Gamma, fcc epsilon, hcp (Co, Cr, ...)	Gamma, fcc (Fe, Ni, Cr, ...)
Geometrically close-packed (GCP)	Gamma, ord. fcc, AB_3 (Ni, Co, Fe, Cr, ...)$_3$(Al, Ti) eta, hep, AB_3 $(Ni)_3Ti$	Gamma' (Ni, CO)$_3$(Al, Ti) eta Ni_3Ti Co_3Ti, ord.fcc Co_3Ti, hcp Co_3W, Co_3Mo, ord. hcp Co_3Ta, ord.cub. Co_3Ta, rhombohedr. CoAl, ord. cub.	Gamma' (Ni, ...)$_3$(Al, Ti) gamma'', ord. bct, AB_3 Ni_3Nb eta Ni_3Ti Delta, orthorhomb., AB_3 Ni_3Nb Beta, bcc, AB Ni (Al, Ti)
Topologically close-packed (TCP)	Sigma, bct, A_xB_y (Cr, Mo) (Fe, Ni) Laves, hcp, A_2B (Fe, Cr, Mn, Si)$_2$ (Mo, Ti, Nb) mu, rhombohedr., A_6B_7 (Mo, W)$_6$(Fe, Co)$_7$ G, fcc, A_6B_{23}	Sigma (Co, Ni) (Cr, Mo, W) Laves Co_2Ta Mu Co_7W_6 R, rhombohedr.	Sigma Cr (Fe, Ni) Laves Fe_2Mo Mu (Co, Ni)$_7$(Cr, W)$_6$ chi, fcc, $Me_{>18}C$

Category			
Carbides	Hf$_6$Ni$_8$Al$_{15}$	(Co, Cr, W, Fe)	(Cr, Fe, Mo)$_{18}$C
		pi	G
		(Co, Ni, Cr, W)C	(Ni, Ti, Si, Co, Fe)
	MC, fcc	MC	MC
	(Ti, Mo, Nb, Ta, W)C	(Ta, Nb, Zr)C	
	M$_7$C$_3$, trigonal	M$_3$C$_2$, rhombic	M$_7$C$_3$
	Cr$_7$C$_3$	Cr$_3$C$_2$	
	M$_{23}$C$_6$, fcc	M$_2$C, hex.	M$_{23}$C$_6$
	(Cr, Mo, Co, W, Nb)$_{23}$C$_6$	Cr$_2$C	M$_6$C
	M$_6$C, M$_3$M$_3$C, M$_4$M$_2$C	M$_7$C$_3$	
	(Ni, Co)$_4$(Mo, W)$_2$C	Cr$_7$C$_3$	
		M$_{23}$C$_6$	
		(Cr, Co, W, Ta)$_{23}$C$_6$	
		M$_6$C, M$_3$M$_3$C, M$_4$M$_2$C	
		(Cr, Ni, Co)$_4$(W, Ta)$_2$C	
Borides	M$_3$B$_2$, tetragonal	?	?
	(Mo, Ti, Cr, Ni)$_3$B$_2$		
	M$_{23}$(C, B)$_6$, fcc		
Nitrides	M(C, N), fcc	M(C, N)	M(C, N)
	Ti(C, N)	(Ta, Zr)(C, N)	MN$_2$
	M$_{23}$(C, N)$_6$		Cr$_2$N
Others	Ni$_x$(MoCr)$_y$(C, Si)	Ni$_5$Zr	M(C, N, P)
	Ti$_4$C$_2$S$_2$		(M, P)$_{23}$C$_6$
	ZrS$_x$		Cr$_3$Ni$_2$Si
			alpha', bcc
			(Cr, Fe)
			Ti$_4$C$_2$S$_2$

date:

(1) Si_3N_4–SiO_2–MgO–CaO system. A mean joining strength of 160 MPa was obtained by joining Si_3N_4 and Si_3N_4 at 1350–1600°C for 1 h under a pressure of 15 MPa.[93] It is believed that the joining materials, Ca and Mg, undergo diffusion from the joining layer to the inside of the parent material.

(2) MgO–Al_2O_3–SiO_2 system. Si_3N_4 and Si_3N_4 were joined at 1575–1650°C for 30–60 min under <1·5 MPa, and a joining strength of ~460 MPa was obtained. The joining was due to a reduction of the Si_3N_4 interface energy and a promotion of wetting by the formation of $SiO_2(l)$ + $Si_3N_4 = 2Si_2N_2O$.[18,19]

(3) MgO–Al_2O_3–SiO_2 system. Si_3N_4 and Si_3N_4 were joined at 1850–1925 K for 30–60 min, and a four-point bending strength of ~460 MPa was obtained at room temperature. At high temperatures, however, the evaporation of Mg and SiO causes a remarkable drop in joining strength, making this system inappropriate for use at high temperatures.[20]

(4) Y_2O_3–La_2O_3–MgO–Si_3N_4 system. A room-temperature joining strength of 500 MPa was obtained at 1450–1700°C for 30 min in an N_2 atmosphere at 25 MPa. Separation was shown at 1000°C.[94]

(5) MO–Al_2O_3–SiO_2 system (M: transition metal ions). Si_3N_4 and Si_3N_4 were joined at 1500–1550°C for 30–60 min in a pressured N_2 atmosphere, and a room-temperature strength of 250 MPa was obtained. Moreover, the high-temperature joining strength at 900°C was virtually unchanged from the room-temperature strength.[23]

(6) CaO–SiO_2–(Y_2O_3) system. A room-temperature strength of 650 MPa was obtained under the same conditions as described in (5) above.[24]

11.8 SUMMARY

Many topics have been left out of this chapter due to lack of space, and the author hopes to present these in their entirety on a future occasion. In addition, an analysis of generated defects using non-destructive testing has been conducted in an attempt to improve Si_3N_4–metal joint reliability, but lack of space prevents the presentation of this work also.

REFERENCES

1. Katz, R. N., Applications of nitrogen ceramics-gas turbines: US National Turbines. In *Progress in Nitrogen Ceramics*, ed. F. L. Riley. Martinus Nijhoff, Leyden, 1977, p. 643.
2. Burke, J. J., Gorum, A. E. & Katz, R. N., (eds), *Ceramics for High Performance Applications*. Metals and Ceramics Information Center, Columbus, Ohio, 1974.
3. Burke, J. J., Gorum, A. E. & Katz, R. N., (eds), *Ceramics for High Performance Applications* II. Brook Hill Publ. Co., Chestnut Hill, 1974.
4. Lenoe, E. M., Katz, R. N. & Burke, J. J., (eds), *Ceramics for High Performance Applications* III. Plenum Press, New York, 1983.
5. Vincenzini, P., (ed.), *Energy and Ceramics*. Material Science Monographs, 6. Elsevier, Amsterdam, 1980.
6. Riley, F. L., (ed.), *Progress in Nitrogen Ceramics*. Martinus Nijhoff, Leyden, 1983.
7. Coe, R. F., British Patent 1,339,541 (5 December, 1973).
8. Wilks, R. S., British Patent 1,417,169 (10 December, 1975).
9. Kaba, T., Shimada, M. & Koizumi, M., *Communs Am. Ceram. Soc.*, **66** (1983), C-135.
10. Becher, P. F. & Halen, S. A., *Am. Ceram. Soc. Bull.*, **58** (1979), 582.
11. Wicker, A., Darbon, P. & Grivon, F., In *Proceedings of International Symposium on Ceramic Components for Engines*, Japan, 1983, p. 716.
12. Suganuma, K., Okamoto, T., Shimada, M. & Koizumi, M., Comment on A treatment of inelastic deformation around a crack tip due to microcracking. *Communs Am. Ceram. Soc.*, **66** (1983), C-117.
13. Elssner, G., Diem, W. & Wallace, J., Microstructure and mechanical properties of metal-to-ceramic and ceramic-to-ceramic joints. In *Surfaces and Interfaces in Ceramics and Ceramic–Metal Systems*, Vol. 14, ed. J. A. Pask & A. G. Evans. Plenum Press, New York, 1981, p. 629.
14. Goodyear, M. U. & Ezis, A., In *Proceedings of 4th Army Materials Technology Conferences*, ed. J. J. Burke, A. E. Gorum & A. Terpinian. Metals and Ceramics Information Center, Columbus, Ohio, 1975.
15. Tabata, H., Kanzaki, S. & Nakamura, M., Solid state joining of silicon nitride ceramics. In *Proceedings of International Symposium on Ceramic Components for Engines*, Japan, 1983, p. 387.
16. Tanaka, S., Nishida, K. & Ochiai, T., Surface characteristics of metal bondable silicon nitride ceramics. In *Proceedings of Interna-*

tional Symposium on Ceramic Components for Engines, Japan, 1983, p. 249.
17. Heap, H. R. & Riley, C. C., British Patent 1,310,997 (21 March 1973).
18. Britten, D., Johnson, S. M., Lamoreaux, R. H. & Rowcliffe, D. J., High-temperature chemical phenomena affecting silicon nitride joints. J. Am. Ceram. Soc., **67** (1984), 522.
19. Johnson, S. M. & Rowcliffe, D. J., Mechanical properties of joined silicon nitride. J. Am. Ceram. Soc., **68** (1985), 468.
20. McCarthney, M. L., Sinclair, R. & Loehman, R. E., Silicon nitride joining. J. Am. Ceram. Soc., **68** (1985), 472.
21. Loehman, R. E., Transient liquid phase bonding of silicon nitride ceramics. In *Surfaces and Interfaces in Ceramics and Ceramic–Metal Systems,* Vol. 14, ed. J. A. Pask & A. G. Evans, Plenum Press, New York, 1980, p. 701.
22. Johnson, S. M., Rowcliffe, D. J., Brittain, R. D. & Lamoreaux, R. H., Joining of silicon nitride. Proc. 2nd Intl. Kolloquium über *"Fügen von Keramik, Glas und Metall,"* Bad Nauheim, 27–29 März 1985, p. 109.
23. Iwamoto, N., Umesaki, N. & Haibara, Y., Silicon nitride joining with glass solder in the system $CaO-SiO_2-TiO_2$. *Yogyo-Kyokaishi,* **94**(8) (1986), 1880.
24. Haibara, Y., Iwamoto, N. & Umesaki, N., Si_3N_4 joining with glasses in the system $CaO-SiO_2$. Paper presented at the annual meeting of the J. Am. Ceram. Soc., Pittsburgh, April, 1987.
25. Andrews, E. H., Destruction of silicon nitride whiskers by reaction with metals at high temperatures. J. Mat. Sci., **1** (1966), 377.
26. Andrews, E. H., Bonfield, W., Davies, C. K. L. & Markham, A. J., Silicon nitride–nickel compatibility: the effects of environment. J. Mat. Sci., **7** (1972), 1003.
27. Allen, B. C. & Kingery, W. D., Surface tension and contact angles in some liquid metal-solid ceramic systems at elevated temperatures. Trans. Am. Inst. Mech. Engrs, **215**(2) (1959), 30.
28. Erz, M. & Hennicke, H. W., Druckloses Fügen Von Siliciumnitrid. In *Keramische Komponenten für Fahrzeug-Gas-turbinen III.* Springer-Verlag, Berlin, 1984, p. 587.
29. Ikuhara, Y., Kobayashi, M. & Yoshinaga, H., Joining of reacted-bonding Si_3N_4 with Al. No. 6 Meeting on High-temperature Materials, 30–31, October 1986, p. 31 (in Japanese).
30. Umebayashi, M., Kishi, K., Tani, E., Kobayashi, K., Ito, S. & Nakamura, R., Wettability with various ceramics and Ni–Mo alloy. J. Japan Ceram. Soc., **93** (1985), 61 (in Japanese).
31. Gifkins, R. C., Diffusional creep mechanisms. J. Am. Ceram. Soc., **51** (1968), 69.
32. Raj, R. & Ashby, M. F., On grain boundary sliding and diffusion creep. Metall. Trans., **2** (1971), 1113.
33. Inomata, K., Surface and interface in ceramics (pp. 102–107). Tension appearance in thin films and their diffusional creep (pp. 224–228). Surface Sci., **4**(102) (1983), 223 (in Japanese).

34. Barsoum, M. W. & Ownby, P. D., The effect of oxygen partial pressure on the wetting of SiC, AlN, and Si_3N_4 by Si and a method for calculating the surface energies involved. In *Surfaces and Interfaces in Ceramics and Ceramic–Metal Systems*, Vol. 14, ed. J. Pask & A. G. Evans. Plenum Press, New York, 1980, p. 457.
35. Elssner, G., Diem, W. & Wallace, J. S., Microstructure and mechanical properties of metal-to-ceramic and ceramic-to-ceramic joints. In *Surfaces and Interfaces in Ceramic and Ceramic–Metal Systems*, Vol. 14, ed. J. Pask & A. G. Evans. Plenum Press, New York, 1980, p. 629.
36. Hennicke, H. W. & Müller, A., Joining of silicon nitride respectively silicon carbide in mixture and/or in contact with high melting metals and superalloys. In *Fügen für Keramik, Glas und Metall*, Baden–Baden, 3–5 Dez., p. 104.
37. Mizuhara, H. & Mally, K., Ceramic-to-metal joining with active brazing filler metal. *Weld. J.*, **64**(10) (1985), 27.
38. Hirai, T. & Matsuda, T., Compatibility of Si_3N_4 with metals at high temperatures. *J. Japan High Temp. Soc.*, **3** (1977), 146 (in Japanese).
39. Koide, K., Sugiura, K. & Mori, M., Current silicon nitride refractory. *Ceramics*, **8** (1973), 816 (in Japanese).
40. Collins, J. F. & Gerby, R. W., New refractory uses for silicon nitride reported. *J. Metals*, **7** (1955), 612.
41. Yashinskaya, G. A., *Ogeneupoly*, **30** (2) (1965), 20.
42. Samsonov, G. V., *Handbook of High Temperature Materials, No. 2; Properties Index*. Plenum Press, New York, 1964, p. 307.
43. Lindop, T. W., Silicon nitride—promising material which can be used as a substitute for heat-resisting alloys. *Kinzoku*, **42**(2) (1972), 42.
44. Mehan, R. L. & McKee, D. W., Interaction of metals and alloys with silicon-based ceramics. *J. Mat. Sci.*, **11** (1976), 1009.
45. Bennett, M. J. & Houlton, M. R., The interaction between silicon nitride and several iron, nickel and molybdenum-based alloys. *J. Mat. Sci.*, **14** (1979), 184.
46. Mehan, R. L. & Bolon, R. B., Interaction between silicon carbide and a nickel-based superalloy at elevated temperatures. *J. Mat. Sci.*, **14** (1979), 2471.
47. Darken, L. S. & Gurry, R. W., *Physical Chemistry of Metals*. McGraw-Hill, New York, 1953.
48. Frank, W., Gösele, U., Mehrer, H. & Seeger, A., Diffusion in silicon and germanium. In *Diffusion in Crystalline Solids*, ed. G. E. Murch & A. S. Nowick. Academic Press, New York, 1984, p. 63.
49. Hall, R. N. & Racette, J. H., Diffusion and solubility of copper in extrinsic and intrinsic germanium, silicon, and gallium arsenide. *J. Appl. Phys.*, **35** (1964), 379.
50. Bakhadyrkhanov, M. K., Zainabinov, S. & Khamidov, A., Some characteristics of diffusion and electrotransport of nickel in silicon. *Sov. Phys.—Semicond.*, **14** (1980), 243.

51. Pell, E. M., Diffusion of Li in Si at high T and the isotope effect, *Phys. Rev.*, **119** (1960), 1014, 1222.
52. Weber, E. R., Diffusion rate of Li in Si at low temperatures. *Appl. Phys.*, **A30** (1983), 1.
53. Gösele, U. & Tan, T. Y., The nature of point defects and their influence on diffusion process in silicon at high temperatures. In *Defects in Semiconductors,* ed. J. W. Corbett & S. Mahajan. Elsevier Science Publishers, North-Holland, New York, 1983, p. 45.
54. Newman, R. C. & Wakefield, J., In *Metallurgy of Semiconductor Materials*, Vol. 15, ed. J. B. Schroeder. Wiley-Interscience, New York, 1961, p. 201.
55. Fuller, C. S. & Ditzenberger, J. A., Diffusion of donor and acceptor elements in silicon. *J. Appl. Phys.*, **27** (1956), 544.
56. Hill, C., Summer Course on Device Impact of New Microfabrication Technologies, Heverlee, Belgium, June 1980.
57. Ghoshtagore, R. N., Donor diffusion dynamics in silicon. *Phys. Rev.*, **B3** (1971), 397.
58. Ghoshtagore, R. N., Intrinsic diffusion of boron and phosphorus in silicon free from surface effects. *Phys. Rev.*, **B3** (1971), 389.
59. Hettich, G., Mehrer, H. & Maier, K., *Inst. Phys. Conf. Ser.*, **46** (1979) 500.
60. Mayer, H. J., Mehrer, H. & Maier, K., *Inst. Phys. Conf. Soc.* **31** (1977), 186.
61. Tu, K. N. & Mayer, J. W., Silicide formation. In Thin films—Interdiffusion and Reactions, ed. J. M. Poate, K. N. Tu & J. W. Mayer. Wiley-Interscience, New York, 1978, p. 359.
62. Bolyaks, B. I. & Hsüch, S. Y., Diffusion, solubility and the effect of silver impurities on electrical properties of silicon. *Sov. Phys. Solid State,* **2** (1961), 2383.
63. Wilcox, W. R. & LaChapelle, T. J., Mechanism of gold diffusion into silicon. *J. Appl. Phys.*, **35** (1964), 240.
64. Navon, D. & Chernyshov, V., Retrograde solubility of aluminum in silicon. *J. Appl. Phys.* **28** (1957), 823.
65. Peart, R. F., Self diffusion in intrinsic silicon. *Phys. Stat. Solidi,* **15** (1966), K119.
66. Masters, B. J. & Fairfield, J. M., Arsenic isoconcentration diffusion studies in silicon. *J. Appl. Phys.*, **40** (1969), 2390.
67. Collins, C. B. & Carlson, R. O., Properties of silicon doped with iron or copper. *Phys. Rev.*, **108** (1957), 1409.
68. Carlson, R. O., Properties of silicon doped with manganese. *Phys. Rev.*, **104** (1956), 937.
69. Bonzel, H. P., Diffusion of nickel in silicon. *Phys. Stat. Solidi,* **20** (1967), 493.
70. Baily, R. F. & Mills, T. G., *Semiconducting Silicon*. Electrochemical Society, New York, 1969, p. 481.
71. Murarka, S. P., Transition metal silicides. In *Ann. Rev. Mat. Sci.*, Vol. 13, ed. R. A. Huggins, R. H. Bube & D. A. Vermilyea. Annual Reviews Inc., California, 1983, p. 117.

72. Murarka, S. P., Refractory silicides for integrated circuits. *J. Vac. Sci. Technol.*, **17** (1980), 775.
73. Chu, W. K., Lau, S. S., Mayer, J. W., Müller, H. & Tu, K. N., Implanted noble gas atoms as diffusion markers in silicide formation. *Thin Solid films*, **25** (1975), 393.
74. Ziegler, J. F., Mayer, J. W., Kircher, C. J. & Tu, K. N., Kinetics of the formation of hafnium silicides on silicon. *J. Appl. Phys.*, **44** (1973), 3851.
75. Schutz, R. J. & Testardi, L. R., The formation of vanadium silicides at thin-film interfaces. *J. Appl. Phys.*, **50** (1979), 5773.
76. Christou, A. & Day, H. M., Silicide formation and interdiffusion effects in Si–Ta, SiO_2–Ta and Si–PtSi–Ta thin film structures. *J. Electron. Mat.*, **5** (1976), 1.
77. Baglin, J., Dempsey, J., Hammer, W., d'Heurle, F., Peterson, S. & Serrano, C., The formation of silicides in Mo–W bilayer films on Si substrates: A marker experiment. *J. Electron. Mat.*, **8** (1979) p. 641.
78. Lau, S. S., Feng, S.-Y., Olowolafe, J. O. & Nicolet, M.-A., Iron silicide thin film formation at low temperatures. *Thin Solid films*, **25** (1975), 416.
79. van Grup, G. J., Sigurd, D. & van der Weg, W. F., Tungsten as a marker in thin-film diffusion studies. *Appl. Phys. Lett.*, **29** (1976), 4301.
80. Tu, K. N., Chu, W. K. & Mayer, J. W., Structure and growth kinetics of Ni_2Si on silicon. *Thin Solid Films*, **25** (1975), 4005.
81. Chu, W. K., Krautle, H., Mayer, J. W., Muller, H., Nicolet, M.-A. & Tu, K. N., Identification of the dominant diffusing species in silicide formation. *Appl. Phys. Lett.*, **25** (1975), 454.
82. Canali, C., Catellani, F., Prudenziani, M., Wadlin, N. H. & Evans, C. A. Jr., Pt_2Si and PtSi formation with high-purity Pt thin films. *Appl. Phys. Lett.*, **31** (1977), 43.
83. Canali, C., Majini, G., Ottaviani, G. & Cellotti, G., Phase diagrams and metal-rich silicide formation. *J. Appl. Phys.*, **50** (1979), 255.
84. Suphet, J. P., Silicides, germanides, stannides. In *Crystal Chemistry and Semiconduction in Transition Metal; Binary Compounds.* Academic Press, New York, 1971, p. 95.
85. Nowotny, H., Crystal chemistry of transition element defect silicides and related compounds. In *The Chemistry of Extended Defects in Non-metallic solids,* ed. L. Eyring & M. O'Keefe, North-Holland, Amsterdam, 1970, p. 223.
86. Irvine, K. J., *J. Int. Applics Cobalt*, Brussels, (1964) p. 286.
87. Betteridge, W., Cobalt containing steels. In *Cobalt and Its Alloys*, Ellis Horwood, Chichester, 1982, p. 99.
88. Tien, J. K., Howson, T. E., Chen, G. L. & Xie, X. S., Cobalt availability and superalloys. *J. Metals*, **32**(10) (1980), 12.
89. Jarrett, R. N. & Tien, J. K., Effects of cobalt on structure, microchemistry and properties of a wrought nickel-base superalloy. *Metall. Trans.*, **13A**(6) (1982), 1021.
90. Tien, J. K. & Jarrett, R. N., Effects of cobalt in nickel-base

superalloys. In *High Temperature Alloys for Gas Turbines,* ed. R. Brunetaud, D. Coutsouradis, T. B. Gibbons, Y. Lindblom, D. B. Meadowcroft & R. Stickler. D. Reidel, Dordrecht, 1982, p. 423.
91. Betteridge, W., *Nickel and Its Alloys.* Ellis Horwood, Chichester, 1984, p. 84.
92. Stickler, R., Phase stability in superalloys. In *High-temperature Materials in Gas Turbines,* ed. P. R. Sahm & M. O. Spiedel. Elsevier, Amsterdam, 1974, p. 115.
93. Owada, Y. & Kobayashi, K., Joining of silicon nitride ceramics. *J. Japan Ceram. Soc.,* **92** (1984), 693 (in Japanese).
94. Tamatoshi, N., Yagi, A., Ebata, Y. & Hihata, Y., Joining between silicon nitride ceramics with aid of Y_2O_3-La_2O_3-MgO-Si_3N_4 mixtures. *J. Japan Ceram. Soc.,* **93** (1985), 154 (in Japanese).

Index

Al_2O_3, 66, 74, 82, 94, 95, 103, 109, 112, 113, 115
AlN, 66, 74, 82
Alumina. See Al_2O_3
Aluminum, Si_3N_4 reaction with, 135, 137

Baking process, 14–15

$CaO–SiO_2–(Y_2O_3)$ system, 150
Carbothermal reduction, 59–69
   crushing/grinding, 65
   decarbonization, 65
   powder
      characterisics, 66–8
      properties, 60
   raw materials and mixing, 60–4
   synthesis conditions, 64–5
   technical elements, 61
Copper, Si_3N_4 reaction with, 136, 140
Crystal
   growth, 61, 78
   structure, 54
   whiskers. See Whiskers
Crystalline phase, 9
Crystallite size, 54–5, 100
Crystallization, 51, 77, 97

Direct nitriding method, 14–17, 27–8, 108
Dissolution after sintering, 10–11

Fabrication, 44–7
Flock, 19–20
FRM (fiber-reinforced metal), 74
FRP (fiber-reinforced plastic), 74

Glassy phase, 9
Grain
   boundaries, materials remaining at, 9–10
   growth, 97
   morphology, 52–4, 75, 101

Imide thermal decomposition, 71–91
   fabrication method, 82–3
   injection molding properties, 86–90
   powder characteristics, 83–5
   process details, 82–3
   synthesis conditions, 76
Injection molding, 86–90

Joining
   methods of, 133–4

Index

Joining—*contd.*
 oxide solder in, 147–50
 research, 132–4
 Si_3N_4–Si_3N_4, 147–50
 silicide formation, 146
 steels
  Co doping, effect of, 146
  Ni and Co diffusion, effect of, 147
  Ni doping, effect of, 147

Melting point, 7
Metallic silicon method, 43–7, 52, 55, 57, 118
MgO, 9, 82
MgO–Al_2O_3–SiO_2 system, 150
MO–Al_2O_3–SiO_2 system, 150
Molybdenum, Si_3N_4 reaction with, 137, 140

Needle crystals, 43
NH_3:$SiCl_4$ ratio, 51, 96
NH_4Cl, removal of, 51
Nickel, Si_3N_4 reaction with, 135
Nickel–30w/t molybdenum, Si_3N_4 reaction with, 138
Nitride ingots, 47
Nitriding
 furnace, 28
 patterns, 47
 reaction, 47
NKK SIN powders, 25–37
 gas-pressure sintering, 33
 molding properties, 32
 normal-pressure sintering, 32–3
 powder characteristics, 28–30
 production process, 27–8
 sintering properties, 32
 slurry characteristics, 30–2
 special sintering, 36

Onoda Cement powders, 13–23
 fabrication, 14–17
 molding properties, 21–2
 raw material powders, 20–2
 sintering, characteristics required for, 13

Oxygen content, 55, 101

Particle size
 adjustment, 16
 distribution, 42, 43, 54
Plasma CVD method, 93–106
 apparatus, 95
 calcining temperature and specific surface area, relation between, 99
 experimental procedure, 94–6
 firing temperature, effect of, 101
 firing temperature and crystallization, relation between, 99–101
 powder
  characteristics, 101
  evaluation, 96
  properties, 97
 reaction conditions, 95
 sintered bodies, 96, 104
 yield, 96–7
Powder characteristics
 measuring, methods of, 108–9
 overview of, 110
 sintering, required for 6–7
 sintering behavior, influence on, 107–16
Powders
 characterization of, 41
 commercial, 42, 43
 high-α-content, 14
 properties of, 48
 synthetic process, 42
precursors, synthesis of, 49–51
Preferential growth orientations, 121
Purity, 72
 adjustment of, 16–17

Runaway reaction, 47

Sample preparation, 18
Sialon, 11
β-SiAlON, 74–6
β-SiC, 74, 76

Si_3N_4
  amorphous, 74
  basic characteristics of, 132
  comparison of powders,
    characteristics of, 51–7
  compatibility of, 141
  properties of, 25, 40, 81–2,
    93–4, 131–2
  raw material powders, 17–20,
    40–3, 48–51
  synthesis of, 44–7
  thermodynamics, 2–7
α-Si_3N_4, 2, 4–6, 59–69, 108, 119
β-Si_3N_4, 2, 4–6, 11, 41, 119
Si_3N_4–SiO_2–MgO–CaO system,
    150
Si–N–O system, phase diagram,
    2–4
$Si_{23}N_{30}O$, 2
α-Si_3Na_4, 16
α-Si_3Ni_4, 72
Silica reduction nitriding synthesis.
    See Carbothermal reduction
Silicide
  formation, 141–6
    dominant diffusing elements
      during, 145
    energy, 143
    thermal stability, 146
Silicon
  diffusion of metals in, 143
  impurities contained in, 44–7
  Ni, Co, Fe, and V, solubility in,
    141–3
  refining, 44–7
  Si_3N_4 reaction with, 138
Silicon chloride and ammonia
    method, 48–52
Silicon diimide ($Si(NH)_2$), 81–91
Silicon halide method, 119
Silicon nitride. See Si_3N_4
Sintering
  dissolution after, 10–11
  hot-pressed, 36
  isostatically hot-pressed, 36
  powder characteristics, influence
    of, 107–16
  properties of, 112–15
  tests, 109

Sintering aids, 7–11, 17, 32, 66,
    81, 94, 96, 103, 109, 112
  characteristics of, 7–8
  classification of, 9–11
  see also specific sintering aids
Stability, 2–4, 5
Stacking faults, 6

Tantalum, Si_3N_4 reaction with, 137
Thermal decomposition of
    $Si(NH)_2$. See Silicon
    diimide
Tin, Si_3N_4 reaction with, 135–6
Tin–titanium alloy, Si_3N_4 reaction
    with, 135–6
Tungsten, Si_3N_4 reaction with,
    137, 140

Vacuum hot-pressing, 125
Vapor–liquid–solid (VLS)
    mechanism, 78, 119
Vapor-phase method. See Plasma
    CVD method
Viscosity measurement, 18–19

Whisker
  additive, 63
  growth, 61, 63
Whisker-reinforced ceramics
    (WRC), 127–9
Whisker-reinforced metals
    (WRM), 123–6
Whisker-reinforced plastics
    (WRP), 126–7
Whiskers, 74, 77–9, 117–30
  applications of, 123–9
  fabrication and characteristics
    of, 118–23
  manufacturing device, 118
  synthesis methods, 118

X-ray diffraction
  analysis, 97, 100
  patterns, 75, 108

Index

Y_2O_3, 9, 66, 82, 94, 95, 103, 109, 112, 113, 115, 134
Y_2O_3–Al_2O_3 system, 32, 33, 147
Y_2O_3–La_2O_3–MgO–Si_3N_4 system, 150

Zirconium, Si_3N_4 reaction with, 138–40
ZrO_2, 134
$ZrSiO_4$, 134

STAFFORDSHIRE
COUNTY REFERENCE
LIBRARY
HANLEY
STOKE-ON-TRENT